TRANSFORMATION IN THE AIR

A REVIEW OF THE FAA'S CERTIFICATION RESEARCH PLAN

Committee to Review the Federal Aviation Administration Research Plan on Certification of New Technologies into the National Airspace System

Aeronautics and Space Engineering Board

Division on Engineering and Physical Sciences

NATIONAL RESEARCH COUNCIL
OF THE NATIONAL ACADEMIES

I0892163

THE NATIONAL ACADEMIES PRESS
Washington, D.C.
www.nap.edu

THE NATIONAL ACADEMIES PRESS **500 Fifth Street, NW** **Washington, DC 20001**

NOTICE: The project that is the subject of this report was approved by the Governing Board of the National Research Council, whose members are drawn from the councils of the National Academy of Sciences, the National Academy of Engineering, and the Institute of Medicine. The members of the committee responsible for the report were chosen for their special competences and with regard for appropriate balance.

This study is based on work supported by Agreement DTFAWA-14-A-80003 between the National Academy of Sciences and the Department of Transportation/Federal Aviation Administration. Any opinions, findings, conclusions, or recommendations expressed in this publication are those of the author(s) and do not necessarily reflect the views of the agency that provided support for the project.

International Standard Book Number-13: 978-0-309-37460-6
International Standard Book Number-10: 0-309-37460-X

Cover: Design by Tim Warchocki.

Copies of this report are available free of charge from

Aeronautics and Space Engineering Board
National Research Council
The Keck Center of the National Academies
500 Fifth Street, NW
Washington, DC 20001

Additional copies of this report are available from the National Academies Press, 500 Fifth Street, NW, Keck 360, Washington, DC 20001; (800) 624-6242 or (202) 334-3313; http://www.nap.edu.

Copyright 2015 by the National Academy of Sciences. All rights reserved.

Printed in the United States of America

THE NATIONAL ACADEMIES

Advisers to the Nation on Science, Engineering, and Medicine

The **National Academy of Sciences** is a private, nonprofit, self-perpetuating society of distinguished scholars engaged in scientific and engineering research, dedicated to the furtherance of science and technology and to their use for the general welfare. Upon the authority of the charter granted to it by the Congress in 1863, the Academy has a mandate that requires it to advise the federal government on scientific and technical matters. Dr. Ralph J. Cicerone is president of the National Academy of Sciences.

The **National Academy of Engineering** was established in 1964, under the charter of the National Academy of Sciences, as a parallel organization of outstanding engineers. It is autonomous in its administration and in the selection of its members, sharing with the National Academy of Sciences the responsibility for advising the federal government. The National Academy of Engineering also sponsors engineering programs aimed at meeting national needs, encourages education and research, and recognizes the superior achievements of engineers. Dr. C. D. Mote, Jr., is president of the National Academy of Engineering.

The **Institute of Medicine** was established in 1970 by the National Academy of Sciences to secure the services of eminent members of appropriate professions in the examination of policy matters pertaining to the health of the public. The Institute acts under the responsibility given to the National Academy of Sciences by its congressional charter to be an adviser to the federal government and, upon its own initiative, to identify issues of medical care, research, and education. Dr. Victor J. Dzau is president of the Institute of Medicine.

The **National Research Council** was organized by the National Academy of Sciences in 1916 to associate the broad community of science and technology with the Academy's purposes of furthering knowledge and advising the federal government. Functioning in accordance with general policies determined by the Academy, the Council has become the principal operating agency of both the National Academy of Sciences and the National Academy of Engineering in providing services to the government, the public, and the scientific and engineering communities. The Council is administered jointly by both Academies and the Institute of Medicine. Dr. Ralph J. Cicerone and Dr. C. D. Mote, Jr., are chair and vice chair, respectively, of the National Research Council.

www.nationalacademies.org

OTHER RECENT REPORTS OF THE AERONAUTICS AND SPACE ENGINEERING BOARD

3D Printing in Space (Aeronautics and Space Engineering Board [ASEB], 2014)
Autonomy Research for Civil Aviation: Toward a New Era of Flight (ASEB, 2014)
Pathways to Exploration: Rationales and Approaches for a U.S. Program of Human Space Exploration (ASEB with the Space Studies Board [SSB], 2014)

Solar and Space Physics: A Science for a Technological Society (SSB with ASEB, 2013)
Final Report of the Committee for the Review of Proposals to Ohio's Third Frontier Program, 2012-2013: Innovation Platform Program 2013 (ASEB, 2013)

Continuing Kepler's Quest: Assessing Air Force Space Command's Astrodynamics Standards (ASEB, 2012)
NASA Space Technology Roadmaps and Priorities: Restoring NASA's Technological Edge and Paving the Way for a New Era in Space (ASEB, 2012)
NASA's Strategic Direction and the Need for a National Consensus (Division on Engineering and Physical Sciences, 2012)
Recapturing NASA's Aeronautics Flight Research Capabilities (ASEB, 2012)
Reusable Booster System: Review and Assessment (ASEB, 2012)

Final Report of the Committee to Review Proposals to the 2011 Ohio Third Frontier Wright Projects Program (OTF WPP) (ASEB, 2011)
Limiting Future Collision Risk to Spacecraft: An Assessment of NASA's Meteoroid and Orbital Debris Programs (ASEB, 2011)
Recapturing a Future for Space Exploration: Life and Physical Sciences Research for a New Era (SSB with ASEB, 2011)

Advancing Aeronautical Safety: A Review of NASA's Aviation Safety-Related Research Programs (ASEB, 2010)
Capabilities for the Future: An Assessment of NASA Laboratories for Basic Research (Laboratory Assessments Board with ASEB, 2010)
Defending Planet Earth: Near-Earth-Object Surveys and Hazard Mitigation Strategies: Final Report (SSB with ASEB, 2010)
Final Report of the Committee to Review Proposals to the 2010 Ohio Third Frontier (OTF) Wright Projects Program (WPP) (ASEB, 2010)

Limited copies of ASEB reports are available free of charge from

Aeronautics and Space Engineering Board
National Research Council
The Keck Center of the National Academies
500 Fifth Street, NW, Washington, DC 20001
(202) 334-2858/aseb@nas.edu
www.nationalacademies.org/aseb.html

COMMITTEE TO REVIEW THE FEDERAL AVIATION ADMINISTRATION RESEARCH PLAN ON CERTIFICATION OF NEW TECHNOLOGIES INTO THE NATIONAL AIRSPACE SYSTEM

S. MICHAEL HUDSON, I Power Energy Systems, *Co-Chair*
WILLIAM S. LEBER, JR., PASSUR Aerospace, *Co-Chair*
JANDRIA S. ALEXANDER, The Aerospace Corporation
STEVEN J. BROWN, National Business Aviation Association
VICTORIA COX, Victoria Cox Solutions, LLC
JOSEPH M. DEL BALZO, JDA Aviation Technology Solutions
R. JOHN HANSMAN, JR., Massachusetts Institute of Technology
AMY R. PRITCHETT, Georgia Institute of Technology
AGAM N. SINHA, ANS Aviation International, LLC
EDMOND L. SOLIDAY, Indiana General Assembly
RAYMOND VALEIKA, Delta Air Lines (retired)
EDWARD L. WRIGHT, University of California, Los Angeles

Staff

DWAYNE DAY, Study Director, Aeronautics and Space Engineering Board
MICHAEL H. MOLONEY, Director, Aeronautics and Space Engineering Board and Space Studies Board
ANDREA REBHOLZ, Program Coordinator, Aeronautics and Space Engineering Board

AERONAUTICS AND SPACE ENGINEERING BOARD

LESTER LYLES, The Lyles Group, *Chair*
PATRICIA GRACE SMITH, Patti Grace Smith Consulting, LLC, *Vice Chair*
ARNOLD D. ALDRICH, Aerospace Consultant, Vienna, Virginia
ELLA M. ATKINS, University of Michigan
STEVEN J. BATTEL, Battel Engineering
MEYER J. BENZAKEIN, The Ohio State University
BRIAN J. CANTWELL, Stanford University
ELIZABETH R. CANTWELL, Lawrence Livermore National Laboratory
EILEEN M. COLLINS, Space Presentations, LLC
MICHAEL P. DELANEY, Boeing Commercial Airplane Group
VIJAY K. DHIR, University of California, Los Angeles
EARL H. DOWELL, Duke University
ALAN H. EPSTEIN, Pratt & Whitney
KAREN FEIGH, Georgia Institute of Technology
PERETZ P. FRIEDMANN, University of Michigan
MARK J. LEWIS, IDA Science and Technology Policy Institute
RICHARD McKINNEY, Independent Consultant
JOHN M. OLSON, Sierra Nevada Corporation
HELEN L. REED, Texas A&M University
AGAM N. SINHA, ANS Aviation International, LLC
ALAN M. TITLE, Lockheed Martin Advanced Technology Center
DAVID M. VAN WIE, Johns Hopkins University Applied Physics Laboratory

MICHAEL H. MOLONEY, Director
CARMELA J. CHAMBERLAIN, Administrative Coordinator
TANJA PILZAK, Manager, Program Operations
CELESTE A. NAYLOR, Information Management Associate
MEG A. KNEMEYER, Financial Officer
SANDRA WILSON, Financial Assistant

Preface

The FAA Modernization and Reform Act of 2012 required the Federal Aviation Administration (FAA) to develop a research plan for the certification of new technologies into the National Airspace System and to have the National Research Council (NRC) review that plan. The FAA has produced various internal planning documents both to guide its research and to determine how it will introduce new technologies, but these are generally not publicly distributed. Starting in the latter 1990s, the FAA began a program named NextGen (Next Generation Air Transportation System) to introduce many new technologies both on the ground and in aircraft to improve the tracking of aircraft and operations. NextGen is the primary program for introducing new technologies in the National Airspace System, but Congress did not explicitly call for a NextGen research plan in the 2012 act. In February 2014, the FAA Office of NextGen produced *Research Plan: Methods and Procedures to Improve Confidence in and Timeliness of Certification of New Technologies Into the National Airspace System*[1] in response to the 2012 act.

In response to an FAA request, the NRC established the Committee to Review the Federal Aviation Administration Research Plan on Certification of New Technologies into the National Airspace System in late summer 2014 to review the research plan. The committee was co-chaired by Michael Hudson and William Leber and consisted of 10 additional members. The membership's expertise included familiarity with FAA management, regulations and operations, commercial, general and business aviation, and relevant areas of research such as software and human factors. The committee met three times: November 19-21, 2014, in Washington, D.C.; January 21-23, 2015, in Irvine, California; and March 3-6, 2015, in Washington, D.C.

The committee's statement of task was as follows:

> An ad hoc committee under the auspices of the National Research Council will conduct an assessment of the Federal Aviation Administration's plan for research on methods and procedures to improve both confidence in and the timeliness of certification of new technologies for their introduction into the National Airspace System (NAS).
>
> The FAA research plan, which will be publicly available, focuses on air-ground and ground-based system approval. The research focus areas in the plan include:
>
> - Systems engineering
> - Requirements development

[1] FAA, *Research Plan: Methods and Procedures to Improve Confidence in and Timeliness of Certification of New Technologies Into the National Airspace System,* Final, Office of NextGen, Washington, D.C., February 2014; reprinted in Appendix A.

- Human factors
- Software development
- Verification, validation, and testing
- Software design assurance
- Training
- Operational evaluation

The committee will:

- Provide an analysis of the FAA's proposed research, as described in the research plan, focused on how implementing the plan may improve both timeliness and confidence of certifying new technologies into the NAS as requested in the FAA reauthorization section 905;
- Identify for each focus area any specific actions that may have been successful in similar activities within and external to the FAA;
- Provide an assessment of planned research outputs in meeting the wording in the 2012 FAA Modernization and Reform Act "to conduct research on methods and procedures to improve both confidence in and timeliness of certification of new technologies for their introduction into the national airspace system";
- Comment on general readability and clarity of presentation of the research plan; and
- Provide comments on or recommendations on other issues determined by the committee.

A key aspect of the committee's efforts was determining how the research plan interacted with other FAA planning documents and how it might guide future research. The plan is a relatively short document (10 pages), and in this report the committee offers a number of findings and recommendations about how the FAA may produce a better research plan in the future that may be more representative of what lawmakers were seeking when they called for it in 2012.

Acknowledgment of Reviewers

This report has been reviewed in draft form by individuals chosen for their diverse perspectives and technical expertise, in accordance with procedures approved by the National Research Council's (NRC's) Report Review Committee. The purpose of this independent review is to provide candid and critical comments that will assist the institution in making its published report as sound as possible and to ensure that the report meets institutional standards for objectivity, evidence, and responsiveness to the study charge. The review comments and draft manuscript remain confidential to protect the integrity of the deliberative process. We wish to thank the following individuals for their review of this report:

Jay Apt, Carnegie Mellon University,
Bruce J. Holmes, NextGen AeroSciences, LLC,
Butler Lampson, Microsoft Corporation,
John K. Lauber, Independent Consultant, Vaughn, Washington,
J. Victor Lebacqz, VICC Associates,
David E. Liddle, U.S. Venture Partners
Dinesh Verma, Stevens Institute of Technology,
David Victor, University of California, San Diego,
Steven Winter, Raytheon Company, and
Andres Zellweger, FAA (ret.)

Although the reviewers listed above have provided many constructive comments and suggestions, they were not asked to endorse the conclusions or recommendations, nor did they see the final draft of the report before its release. The review of this report was overseen by M. Granger Morgan, Carnegie Mellon University. Appointed by the NRC, he was responsible for making certain that an independent examination of this report was carried out in accordance with institutional procedures and that all review comments were carefully considered. Responsibility for the final content of this report rests entirely with the authoring committee and the institution.

Contents

Summary

The Federal Aviation Administration (FAA) is currently undertaking a broad program known as NextGen (Next Generation Air Transportation System) to develop, introduce, and certify new technologies into the National Airspace System. NextGen is a fundamentally transformative change that is being implemented incrementally over a period of many years.[1] Its implementation is analogous to the introduction of radar during the 1950s; radar was also a transformative change in domestic air travel, but it took many years to be fully implemented and for users to adapt and take advantage of its capabilities. Currently, the FAA is putting into place the foundation that provides support for the future building blocks of a fully operational NextGen.

NextGen is a challenging undertaking that includes ground systems, avionics installed in a wide range of aircraft, and procedures to take advantage of the new technology. Not all of these systems or procedures are under the FAA's control, a reality that determines implementation and therefore requires close coordination among the FAA and various stakeholders.[2] NextGen is also best understood as a continuing evolution of systems and procedures, not simply technology systems that have to be developed and deployed. Like all complex systems of systems, the implementation has experienced problems, but has also made important advances, some of which are highlighted in Chapter 3 of this report.

The FAA has continuously updated the National Airspace System for decades, although what is generally referred to as NextGen had its origins in the late 1980s but was really only formalized by the late 1990s. Since that time the conditions of, and justification for, this enormous undertaking have changed. The biggest of these changes are the following:

- The overcrowding of the airspace that resulted in air travel gridlock almost a decade ago has faded with reduced demand for air travel during the economic downturn. Airlines have also responded to high fuel costs by operating fewer aircraft with more passengers on board. As a result, the need for greater airspace capacity in the short term has diminished.

---

[1] This committee was not asked to evaluate NextGen. For a broader perspective on NextGen, see the 2015 National Research Council (NRC) report *A Review of the Next Generation Air Transportation System: Implications and Importance of System Architecture*, The National Academies Press, Washington, D.C.

[2] The term "stakeholders" in the National Airspace System is a broad one, covering many entities. The stakeholders include the airlines, the general aviation community, the U.S. military, manufacturers, and the traveling public. Thus, the FAA's research and its certification processes can affect many different entities in different ways.

- Other concerns such as efficiency, environmental, and noise impacts of aviation have increased during this same period of time.
- The anticipated significant increase in general aviation traffic has not occurred.
- The current safety record of the National Airspace System is at an all-time high. Safety has not deteriorated but has actually improved statistically within the last few years to unprecedented levels. Paradoxically, this high safety rate has resulted in little incentive by those required to invest in the system to meet NextGen's future safety goals while increasing the number of operations in the National Airspace System.

The operators, such as airlines, support an improved system, but the benefits for them are not as clear today as they seemed years earlier, particularly when measured against the significant investments they are asked to make. Thus, there is less incentive for faster implementation of new technologies into the National Airspace System than there was over a decade ago.[3]

While the FAA can be a capable program manager and direct public capital investment, it does not control investment in and implementation by a broad and diverse operator community in necessary technology, training, and other elements required in an integrated plan. This diverse stakeholder community seeks a broad set of differing operational benefits from the FAA. All stakeholders would benefit substantially from an explanation of the end-to-end processes necessary to certify, approve, and implement advanced NextGen capabilities beyond the mid-term (i.e., 5-7 years).

The FAA Modernization and Reform Act of 2012 required the Federal Aviation Administration to develop a research plan for the certification of new technologies into the National Airspace System and to have the National Research Council (NRC) review that plan.[4] The NRC's Committee to Review the Federal Aviation Administration Research Plan on Certification of New Technologies into the National Airspace System reviewed the February 2014 research plan of the FAA Office of NextGen, *Research Plan: Methods and Procedures to Improve Confidence in and Timeliness of Certification of New Technologies Into the National Airspace System,*[5] and determined that the plan lacks detail and specificity and does not provide an effective guide to FAA research over the 5-year term required by the act. The committee concluded that the plan does not meet the requirements of the authorizing legislation. Whereas the plan restates the language from the FAA Modernization and Reform Act of 2012, it lacks the specificity required to generate actionable objectives.[6] It is more of a high-level task plan for incrementally developing over the next 5 years the detailed research plan that the FAA will actually need.

In particular, the February 2014 Research Plan does not demonstrate how the integration of aircraft, ground systems, and procedures will occur. Successfully demonstrating this will create confidence in implementation and attract stakeholder and operator investment, which is vital for success. The committee also was concerned with the lack of detail in the plan on the transition of research into applications through a structured certification approach that results in an approved operational capability. The plan contains a large amount of content on background and scope and assumptions, but very little on product schedule, milestones, and budgeting. In addition, the committee

---

[3] The FAA's Acquisition Management System (AMS), which is discussed in Chapter 1, requires consideration of costs and benefits.

[4] Regarding research on design for certification, the act stated the following:

> Research—Not later than 1 year after the date of enactment of the FAA Modernization and Reform Act of 2012, the Administrator shall conduct research on methods and procedures to improve both confidence in and the timeliness of certification of new technologies for their introduction into the national airspace system.
>
> Research plan—Not later than 6 months after the date of enactment of the FAA Modernization and Reform Act of 2012, the Administrator shall develop a plan for the research under paragraph (1) that contains objectives, proposed tasks, milestones, and a 5-year budgetary profile.
>
> Review—The Administrator shall enter into an arrangement with the National Research Council to conduct an independent review of the plan developed under paragraph (2) and shall provide the results of that review to the Committee on Science, Space, and Technology of the House of Representatives and the Committee on Commerce, Science, and Transportation of the Senate not later than 18 months after the date of enactment of the FAA Modernization and Reform Act of 2012.

[5] FAA, *Research Plan: Methods and Procedures to Improve Confidence in and Timeliness of Certification of New Technologies Into the National Airspace System,* Final, Office of NextGen, Washington, D.C., February 2014.

[6] The February 2014 Research Plan is reprinted in Appendix A of this report. The plan includes a timeline listing the key points in development and approval of the plan. Finalized in February 2014 and approved by FAA senior management in April 2014, it was made available to the NRC in summer 2014, at which point the NRC created a committee to review it and hold meetings to gather additional information. NRC committee meetings occurred in November 2014 and January and March 2015.

believes that one of the missed opportunities of the plan was that it did not explain how much progress has already been made or what is left to be done and does not reference or provide material from the NextGen Strategic Plan[7] or the National Aviation Research Plan.[8]

The plan also uses the term "certification" inconsistently, using the term "approval" interchangeably with "certification" in some cases. A more meaningful milestone than *certification* of the new technology is *the approval of the operational capability of that technology for implementation by National Airspace System users*. The committee used this broader view of the term certification in its assessment.

Assessing the timeliness of the research plan requires understanding the stakeholders, benefits, and control over what gets done in order to certify new technologies for introduction into the National Airspace System, and who does it. The FAA is the central focus of the certification of new technologies, but the FAA does not have control over the stakeholders or over the derivation of the benefits to the stakeholders.

The committee has concluded that it is in the best interests of the FAA and its stakeholders for the FAA to describe and carefully explain the steps that the FAA and aviation stakeholders are taking to expedite the realization of the NextGen capabilities. Thus, there is value to the FAA's producing a comprehensive research plan that explains its research goals and plans for integrating and certifying technology into the National Airspace System. The committee concluded that future FAA research plans, when properly assembled and executed, can play a valuable role in guiding the FAA and stakeholders and explaining progress in certifying new technologies into the National Airspace System. But although a research plan can help, without goals and operational performance-based metrics such as fuel burn, capacity, delays, cancellations, carbon emissions, and other relevant factors, a plan by itself cannot control the pace of implementation of capabilities or the realization of stakeholder operational benefits. These kinds of metrics are found in other FAA documents, but are not present in the February 2014 Research Plan.

RECOMMENDATION: In order to improve confidence in and timeliness of the certification of new technologies and the approval of the new operations they enable in the National Airspace System, the FAA should create a comprehensive research plan that results in a documented approach that provides the full context for its certification and implementation of NextGen, including both ground and air elements, and the plan's relationship to the other activities and procedures required for certification and implementation into the National Airspace System. The current plan does not do this.

Because the February 2014 Research Plan does not include air systems or procedures, it is unlikely that the plan by itself would address all the elements necessary to improve the timeliness and effectiveness of the certification and implementation of technologies into the National Airspace System. The National Airspace System is tightly integrated, and air and ground capabilities are closely linked with each other and with operational procedures. This requires that all three segments be addressed by a research plan. Omitting air systems and procedures from the research plan makes it unresponsive to the request from Congress, in the committee's opinion. In addition, the plan does not reference other agency reports, plans, and resources that inform and frame research to improve confidence in and timeliness of certification of new technologies.

During the course of this study, the committee concluded that Congress and the FAA were not effectively communicating and there was some misunderstanding on both sides. Congress wrote legislation calling for a "research plan," but according to discussions with congressional staff, Congress actually wanted a detailed description of processes, plans, and capabilities to certify new technologies in a timely manner, of which a research plan is only one key part. The FAA responded by producing a research plan that is too narrow in scope and failed to explain all of the other relevant factors in the timely certification of new technologies and progress in implementing them. Better communication between the FAA and Congress prior to production of the February 2014 Research Plan by the FAA would have ameliorated this confusion. However, even taking this communication failure into account,

[7] FAA, *NextGen Strategic Plan*.

[8] FAA, *National Aviation Research Plan*, September 2013, https://www.faa.gov/about/office_org/headquarters_offices/ang/offices/tc/about/campus/faa_host/rdm/media/pdf/2013%20NARP.pdf. The February 2014 Research Plan in Appendix A makes a single cursory reference to the NARP (see p. 34).

the February 2014 Research Plan is still too narrow in scope (for instance, omitting the aircraft segment) to satisfy the requirements of the 2012 act. The committee believes that future improved communication between congressional staff and FAA staff could alleviate concerns and be highly productive.

OBSERVATIONS ON FAA RESEARCH PLAN RESPONSIBILITIES AND IMPLEMENTATION

The complexity of the process of certifying new technologies also leads to unrealistic expectations by the stakeholders. This complexity begins with the significant relationships between initial hardware or process certification and subsequent broad implementation and approval of new operations in the National Airspace System. While it is easy to believe that once a system is certified the job is finished, the reality is that certification of a particular system component is only an important milestone in a larger process of implementation. This implementation diffuses the control not only to within different organizations within the FAA but also to stakeholders, who must also make investments and implement changes within their organizations. Thus, the FAA is not in a position to single-handedly enact or regulate the intended benefits of NextGen implementation, and their regulations may impose costs.

The implementation effort creates issues affecting aircraft modifications, ground-based system upgrades, training at both the FAA and stakeholders, development of new procedures, revision of operational regulations, and so on. During this time of change, human factors considerations are critical to ensure safe operations within a system in flux, and their inclusion in a research plan will be a key part of its success.

It is obvious from this discussion that the challenge is to create a single fully integrated system from an amorphous mass of projects and stakeholders. The February 2014 Research Plan does not provide a plan to research certification and implementation of such a system.

There are many components to an effective research plan. The committee in this report sought to identify several that it believes deserve particular attention, in part because they are often overlooked or underemphasized and in part because they are gaining increasing importance. A well-planned systems engineering approach is essential to the success of the integration and implementation of a complex multi-faceted system of systems like NextGen. Currently, the FAA is almost entirely reliant upon vendors for software assurance, but the plan does not describe how the FAA is conducting a robust software assurance program. Fundamental to these requirements is the development of an enterprise architecture and National Airspace System-level system architecture. These tools are critical for setting the context for all levels of research and for determining which subjects should be investigated as priority and which topics could offer the most benefits in the nearer term.

RECOMMENDATION: The FAA research plan should address software assurance issues associated with complex systems in order to ensure timeliness and confidence in the certification of new technologies into the National Airspace System.

Cybersecurity, including the important issue of data privacy and verification and validation of new systems to ensure that they work, are also vital aspects to the success of FAA systems. Cybersecurity is a critical component of the National Airspace System that needs to be addressed early, continually, and comprehensively across the systems, segments, and procedures. Similarly, verification and validation is a vital but easily overlooked issue, and the committee determined that it deserves specific attention in any research plan because it can create significant implementation delays.

RECOMMENDATION: The FAA research plan should address cybersecurity as an integral part of the National Airspace System.

RECOMMENDATION: The FAA research plan should include as a significant priority the improvement in the use of verification and validation of the overall system. The FAA research plan should demonstrate how the FAA is building upon the significant research on verification and validation

being done by NASA and other government research labs, academia, and international research groups.

Although the FAA has a research budget and research facilities, a substantial amount of research on new air traffic control systems, procedures, and related technologies is performed by NASA. The February 2014 Research Plan does not refer to research conducted by NASA or other organizations that perform relevant research (for instance, the development of new air traffic control technologies or regarding the certification of new technologies) or engage in analogous activities, some of which is discussed in Chapter 3.

RECOMMENDATION: The FAA research plan should benchmark the best practices of other organizations regarding certification that can contribute to the timely implementation of NextGen technologies and coordinate its research with other relevant organizations, particularly NASA.

Chapter 1 of this report addresses the committee's review of the February 2014 Research Plan, explains the components of an effective research plan, and formulates findings and recommendations. Chapter 2 of this report examines specific shortfalls in the plan and offers suggestions on issues to be included in any future plan. Chapter 3 of this report addresses examples of successful NextGen projects as well as the work of other organizations such as the U.S. Air Force and NavCanada with analogous experience. Chapter 4 of this report addresses the committee's information gathering from the FAA and interpretation of activities related to the research plan. Chapter 4 also discusses some of the extensive work the FAA has done to date to implement the incremental NextGen approach.

1

A Review of the FAA Research Plan

In response to the FAA Modernization and Reform Act of 2012, the FAA Office of NextGen prepared a 10-page research plan[1] for the Next Generation Air Transportation System (NextGen) that was completed in February 2014 and approved by FAA management in April 2014. (Because the FAA has other research plans covering other activities, this report will refer to the "February 2014 Research Plan" throughout this report.) In response to the request to review the research plan, the National Research Council created a committee that gathered data from the FAA, congressional staff, industry, and other sources to assist in its review. The committee received briefings by the individual that generated the February 2014 Research Plan and was able to have dialogue with the relevant FAA officials and discuss the management guidance that went into generating it.

The FAA Modernization and Reform Act refers to methods to improve confidence in the certification of new technologies. True confidence requires ownership by the various stakeholders in the application and usage of NextGen capabilities. (Chapter 3 of this report addresses examples of successful projects that have helped stakeholders gain confidence in NextGen implementation.)

The transition of technologies into the National Airspace System and the generation of the associated procedures, regulations, and certification processes is a major challenge for the FAA. One of the problems is that improvements in avionics systems are occurring at a far more rapid pace than the procedures, regulations, and certification processes. Another issue is that avionics systems are becoming relatively cheaper whereas the certification costs are not. When the results of research are handed over for certification, a whole new process begins where the resulting new equipment must be designed, built, and then certified. Different functions of the FAA are required to be engaged in those processes—from airworthiness, to operational specification approval, to training, to certifying new air traffic procedures—and provide the interface with the industries producing the hardware and the operators that use the system.

Certification of the new technology is not as important as the approval of the operational capability of that technology and its ultimate implementation in the National Airspace System. The many stakeholders play a major role, from the airlines and other users buying, installing, training, and using the new capabilities, to the operators of the National Airspace System having sufficient training, procedures, regulations, and policies to take advantage of the technology.

[1] FAA, *Research Plan: Methods and Procedures to Improve Confidence in and Timeliness of Certification of New Technologies Into the National Airspace System,* Final, Office of NextGen, Washington, D.C., February, 2014; reprinted in Appendix A.

The users and supporters of the National Airspace System are a very broad mix of stakeholders from airlines to the military to general aviation, the manufacturers of the air and ground equipment, and multiple labor organizations, all with different and sometimes conflicting interests and expectations. The operation of the National Airspace System affects the lives of people around the world in terms of travel, commerce, and national security. This in turn presents the FAA with a complex and dynamic set of challenges. All of these users and stakeholders have a significant impact on the scope and type of research that the FAA must conduct.

An effective plan for research on methods and procedures to improve both confidence in and the timeliness of certification of new technologies for their introduction into the National Airspace System should capture the strategic and vision-oriented expectations of the entire FAA organization and its stakeholders. The details of the translation of the plan from vision, to objectives, to tasks, to implementation of outcomes and operations in the real world are fundamental to the success of the program. The plan would be a high-level description of the FAA research planning process that includes the following elements:

- A description of the strategic and prescriptive value of a research plan;
- A description of the value of the expected content of such a plan; and
- An explanation of the expected outcomes from executing the plan.

The following goals would be addressed in a comprehensive research plan:

- Enhancing timeliness,
- Improving confidence,
- Adopting a total system perspective,
- Acknowledging user adoption/operational transition,
- Addressing overall approval as well as certification,
- Increasing integrated accountability by researching critical dependencies and defining a clear and achievable outcome, and
- Integrating emerging technologies upfront.

Instead of a high-level description of the FAA research planning process, the committee concludes that the February 2014 Research Plan is more of a high-level *task plan* for incrementally developing a detailed research plan over the next 5 years. In other words, it is *a plan for developing a plan*. It fails to address the full scope of research necessary to meet the direction from Congress.

The research plan states that it represents "the FAA strategy for conducting research on methods and procedures to improve the confidence and timeliness of certification of new technologies." While this objective is well summarized in introductory material in the plan, the plan includes only limited discussion of processes, programs, and procedures to achieve this objective. Also missing is the presentation of an integrated approach or an end-to-end process that would ultimately result in a cohesive, comprehensive, and integrated plan for transitioning certified technologies into actual end-user capabilities in the National Airspace System.

The committee is concerned that the FAA February 2014 Research Plan cannot improve timeliness or be effective for the following reasons:

- It assumes a traditional definition of "research." Research in the traditional sense can take years to produce results.
- The schedule presented fails the timeliness test—witness the time taken just to write this 10-page plan that presents an approach that will deliver a "report" in the last quarter of 2018 that gets the FAA ready to develop an implementation plan. The plan implies that the FAA will be starting to "plan" how to be more timely when the bulk of the NextGen technology has already been delivered. The committee can only conclude that this timeframe is not what Congress had in mind when tasking the FAA with developing a research plan.

- The plan includes neither a flow diagram of tasks and milestones, nor does it have any mapping of a holistic plan that includes both ground and air.
- The plan focuses on system design without addressing facility and service approval processes (Figure 1.1), missing the opportunity to research process improvements that might deliver near- and mid-term capabilities in a timely manner. Any improvements to timely delivery will accrue to capabilities well outside the NextGen mid-term (i.e., 5-7 years).
- The plan does not discuss research on critical dependencies that could increase integrated accountability, such as an illustration of linkages and interdependencies of the approval process elements that are listed in Figure 1.1.
- The plan does not explain that this plan is one part of a larger set of documents that describe the FAA's implementation of NextGen. The plan misses the opportunity to methodically outline the current processes so they can be reviewed for improvement. Such an outline could have been provided by reference, and the holistic view and mapping to the basic objectives of the outline might have identified high-benefit, quick-return improvement topics. Further, it could have addressed the integration of lessons learned into ongoing and future programs.

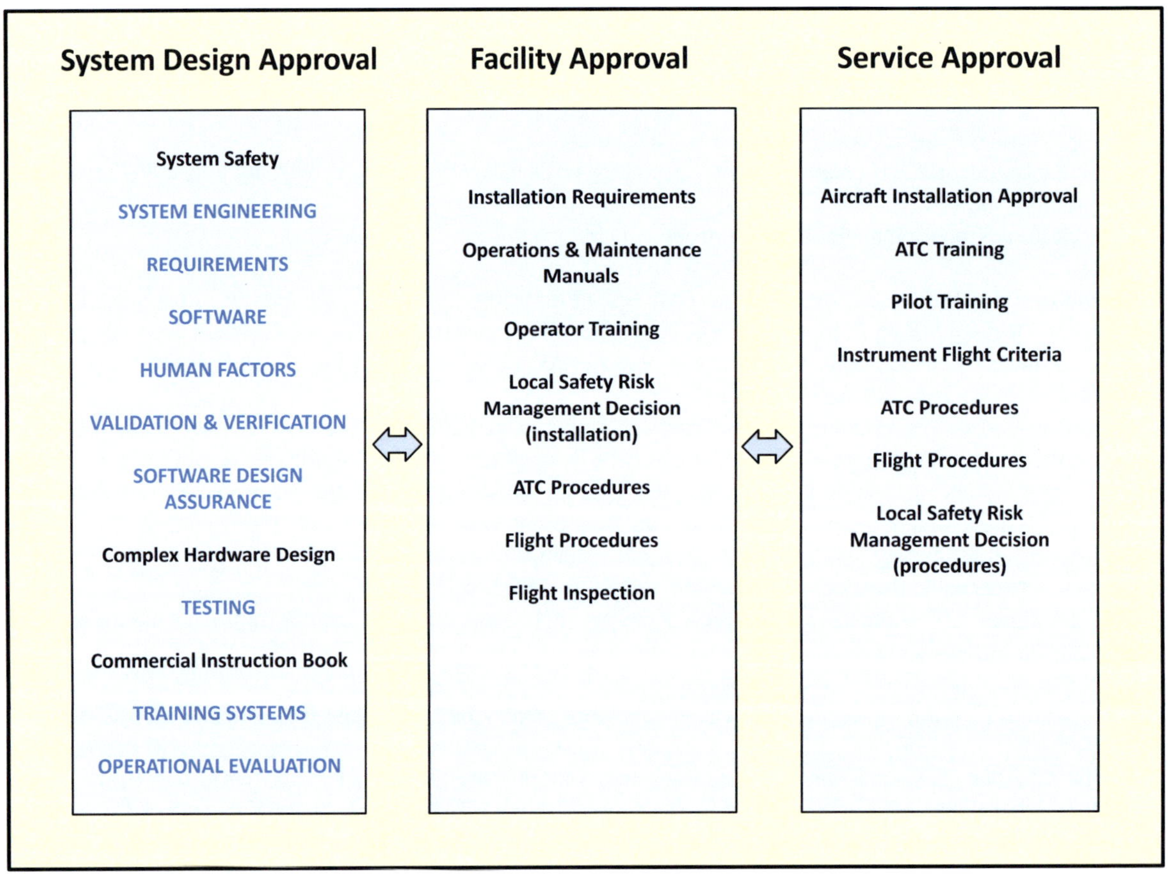

FIGURE 1.1 National Airspace System ground-system approval process overview. SOURCE. Federal Aviation Administration, *Research Plan: Methods and Procedures to Improve Confidence in and Timeliness of Certification of New Technologies Into the National Airspace System,* Final, Office of NextGen, Washington, D.C., February 2014.

- The plan shows no recognition of the implementation of the evolutionary improvements that have been and are currently being made in the National Airspace System and how any proposed new technology may be introduced. The plan does not make clear how the research in this plan is being used to accelerate the use of technology found in existing certified hardware and procedures that might benefit the National Airspace System.
- The plan does not reference other agency reports, plans, and resources that inform and frame research to improve confidence in and timeliness of certification of new technologies. These include the NextGen Implementation Plan,[2] which sets the context and goals for delivery of NextGen capabilities; the National Aviation Research Plan,[3] which details investment of FAA's research budget, some current elements of which are relevant to timeliness and confidence of certification; the National Airspace System Enterprise Architecture,[4] the comprehensive blueprint being used to build NextGen; the NextGen Priorities Joint Implementation Plan,[5] which lays out priorities for implementation through 2018, developed jointly with the NextGen Advisory Committee; and the Navigation Procedures Initial Implementation Plan (NAV Lean) Report,[6] which identifies ways that the FAA will streamline the processes necessary to implement new procedures.
- The plan also does not discuss how its research relates to the responses prepared by the FAA to Sections 215 (Certification Standards and Resources) and 312 (Aircraft Certification Process Review and Reform) of the FAA Modernization and Reform Act of 2012—which address other aspects of certification (aircraft, equipage, procedures). The plan does not indicate how these efforts together will improve timeliness and confidence.
- The plan could enhance the FAA's ability to gain confidence in software and hardware systems through testing and statistical methods, but the relative proof of confidence is proving to the user (and the public) that implementation will produce economic, safety, security, or other benefits. However, the plan does not include these elements:
  - —Details for how validation and verification would occur.
  - —Performance-based metrics (for example, fuel burn, delays, and carbon emissions) common to many other FAA documents that allow stakeholders to determine what the goals are and how they are being achieved.
- Even though the plan itself states, "In preparing this report, we have considered certification process for two basic categories of technologies . . . . (1) those associated with aircraft; and (2) those associated with ground based systems and air traffic control (ATC)," the plan only addresses the second item, ground-based systems, and *not* the first, despite the fact that aircraft systems are a vital, contentious, and complex issue, and procedures certification remains a major bottleneck for implementation. Committee discussions with FAA representatives revealed that aircraft systems were omitted from the February 2014 Research Plan at the direction of FAA management. No explanation was provided for this decision. Not including aircraft systems or procedures makes it unlikely that the plan by itself would address all the elements necessary to improve the timeliness and effectiveness of the certification and implementation of technologies into the National Airspace System. The focus of the plan on only one segment of the National Airspace System is a major deficiency. The tightly integrated nature of the system—whereby the ultimate success of implementing the program depends on the integration of air and ground capabilities with operational procedures—requires that all three segments be addressed by a research plan. Omitting aircraft systems and procedures from the research plan makes the plan unresponsive to the request from Congress.
- The plan's focus on certification of technologies without sufficient integration with and approval of the

---

[2] FAA, *NextGen Implementation Plan*, August 2014, https://www.faa.gov/nextgen/library/media/NextGen_Implementation_Plan_2014.pdf.

[3] FAA, *National Aviation Research Plan*, September 2013, https://www.faa.gov/about/office_org/headquarters_offices/ang/offices/tc/about/campus/faa_host/rdm/media/pdf/2013%20NARP.pdf.

[4] National Airspace System Enterprise Architecture.

[5] FAA, *NextGen Priorities Joint Implementation Plan*, October 2014, http://www.faa.gov/nextgen/media/ng_priorities.pdf.

[6] FAA, *Navigation Procedures Initial Implementation Plan (NAV Lean)*, June 1, 2011, http://www.faa.gov/nextgen/media/SIGNED%20Initial%20NavLean%20Implementation%20Plan%201%20June%202011.pdf.

operations they will enable, does not provide any mechanisms for examining systems and operations that require a close integration of air and ground technologies.

- A major missing element in the plan is a holistic analysis of the FAA's many stakeholders who play a variety of roles in the processes leading to certification and ultimate use of NextGen improvements. The plan does not discuss in depth the ability to organize the stakeholders or direct them with a view to creating timely implementation.
- The plan does not mention efforts at global harmonization, one of the most important reasons for a systemic approach to NextGen. The FAA may be taking actions to assure a systemic approach globally, but the plan does not mention this in the proposed research or in present ongoing FAA efforts.
- The plan also fails to reference Acquisition Management Guidance, as required by the FAA's Acquisition Management System (AMS) policy, as well as an outline of how it will be applied. The AMS provides mandatory requirements for procurement, deployment of products and services, and in-service management of fielded capabilities and, therefore, is directly applicable to the transition of the NextGen technology to operational application.
- The plan does not follow a normal technical reporting format and could have included more information that would have enhanced the public's understanding of and confidence in the FAA's work. In areas where information was provided, more detail would have established greater insight and confidence that the FAA can deliver NextGen in a timely fashion.
- The plan also does not discuss any possible organizational changes to improve efficiency or effectiveness. Although reorganization can create additional problems while sometimes failing to solve others, it may be useful and necessary in limited cases to help clarify lines of authority.

COMPONENTS OF AN EFFECTIVE FAA RESEARCH PLAN

The goals discussed below would normally be addressed as part of a more comprehensive research plan.

Enhance Timeliness

A valid research plan would characterize the steps involved in developing, certifying, and transitioning technology into operation in the National Airspace System. The characterized process would then serve as the basis for defining areas of improvement.

Improve Confidence

Process improvements that lead to immediate realization of benefits will increase confidence in the FAA's ability to fully implement technology modernization. A successful design, even though tested and operationally evaluated, will not deliver benefits until it has transitioned into the operation and users have adopted it. A successful plan would characterize the technical modernization activities from an end-user perspective, for example, citing these programs as enhancements or a gap filler or new capabilities. Including an integrated perspective and presenting the targeted impact or improvements to the user would instill confidence that many well-executed disparate program components are manageable as an integrated program and truly improve the baseline.

Adopt a Total System Perspective

A plan focusing on research aimed at all aspects of approval, including accelerating operational transition, cross-organizational collaboration, and user adoption, would give a total system view that includes integrated testing, validation activities, and cybersecurity in operationally representative environments. These key technologies and others, such as human factors, need to be rigorously addressed. A viable plan would present how technologies will be integrated and addressed throughout the implementation and adoption process. While critical and challenging, integration needs to occur as an integral part of programs and integrated system development without disrupting the overall safety and quality posture.

Global harmonization—the need to harmonize aircraft systems and ground systems as much as possible across the entire world—is one of the most important reasons for a systemic approach to NextGen. Airlines and other operators of intercontinental aircraft cannot be expected to meet multiple and disparate levels of mandates for equipping their airplanes. The European Union's Single European Skies Air Traffic Management Research program (SESAR) and NextGen have similar goals and mandates, and although the FAA may be taking actions to assure a systemic approach globally, a complete research plan would mention these efforts either in proposed research or in present, ongoing FAA efforts.

FAA initiatives aimed at aircraft and aircraft equipage certification would also be noted in a comprehensive research plan. The FAA's initiative focusing on streamlining the implementation of procedures in the National Airspace System with its NAV Lean project, for example, followed the AMS process for implementation as described in the report *Navigation Procedures Initial Implementation Plan (NAV Lean).*[7]

Acknowledge User Adoption/Operational Transition

Operational transition and user adoption were identified as challenges that drive gaps between the FAA's documented descriptions of NextGen and what is being accomplished in a recent comprehensive and independent assessment of NextGen by MITRE.[8] These gaps, which contribute to the different perceptions within the community about the amount of progress the FAA has made on NextGen, need to be addressed.

Address Overall Approval as well as Certification

System certification alone does not guarantee approval for use in all applications, nor does it guarantee user adoption. The committee concludes that a complete plan would go beyond traditional technology-based certification.

Increase Integrated Accountability

An effective research plan would include these two components to increase integrated accountability:

- *Research critical dependencies.* Focusing on understanding critical approval dependencies, intersecting organizational responsibilities, linkages between decision points and decision makers, and barriers to successful operational transition and user adoption can produce an illustration of linkages and interdependencies of the approval process elements that are listed in Figure 1.1.
- *Define a clear and achievable outcome.* Clear articulation of a common understanding of the path to a successful implementation of NextGen capabilities across the stakeholder spectrum could lead to a better ability to streamline processes and to an improved confidence in the FAA's ability to deliver NextGen.

Upfront Integration of Emerging Technologies

A better plan would address methods for integrating emerging technologies early and throughout the system development, certification, and implementation process. An approach might include addressing the integration of the topics cited in the FAA plan, including software assurance, cybersecurity, and human factors, while maintaining comprehensive safety and compliance standards.

---

[7] FAA, *Navigation Procedures Initial Implementation Plan (NAV Lean)*, 2011.

[8] "NextGen Independent Assessment and Recommendations," MITRE Project No. 0214DL01-IF, Center for Advanced Aviation System Development, October 2014.

FINDING: NextGen is a fundamentally transformative change that is being implemented incrementally over a period of years. Currently, the FAA is putting into place the foundation that provides support for the future building blocks of a fully operational NextGen.

FINDING: The February 2014 Research Plan does not meet the requirements of the authorizing legislation. The plan restates the language from the FAA Modernization and Reform Act of 2012, but lacks the specificity required to generate actionable objectives.

FINDING: The February 2014 Research Plan does not demonstrate how integration of aircraft, ground systems, and procedures will occur in the National Airspace System. Successfully demonstrating this will create confidence in implementation and attract stakeholder and operator investment.

FINDING: It is in the best interests of the FAA that it describe and fully explain the steps that the FAA and aviation stakeholders are taking to expedite the realization of the NextGen capabilities. There is value to the FAA producing a comprehensive research plan that explains its research goals and plans for integrating and certifying technology into the National Airspace System. Future FAA research plans, when properly executed, can play a valuable role in guiding the FAA and stakeholders and explaining progress in certifying new technologies into the National Airspace System.

FINDING: Without goals and operational performance-based metrics such as fuel burn, capacity, delays, cancellations, carbon emissions, and other relevant factors, a research plan by itself cannot control the pace of implementation of capabilities or the realization of stakeholder operational benefits. These metrics are found in other FAA documents, but are not reflected in the February 2014 Research Plan.

FINDING: While the FAA can be a capable program manager and direct public capital investment, it does not control investment and implementation by a broad and diverse operator community in necessary technology, training, and other elements required in an integrated plan. This diverse stakeholder community seeks a broad set of differing operational benefits.

FINDING: All stakeholders would benefit substantially from the explanation of the end-to-end processes necessary to certify, approve, and implement advanced NextGen capabilities beyond the mid-term (i.e., 5-7 years).

RECOMMENDATION: In order to improve confidence in and timeliness of the certification of new technologies and the approval of the new operations they enable in the National Airspace System, the FAA should create a comprehensive research plan that results in a documented approach that provides the full context for its certification and implementation of NextGen, including both ground and air elements and the plan's relationship to the other activities and procedures required for certification and implementation into the National Airspace System. The February 2014 Research Plan does not do this.

2

Specific Shortfalls in the February 2014 Research Plan

There are a number of issues that the February 2014 Research Plan[1] addresses only superficially and are in need of more extensive discussion in any future research plan. Many of these issues are both highly complex and interrelated due to the nature and scope of NextGen. NextGen is a large, complex system made up of systems, each of which is a "system of systems." NextGen also exists within a global context, meaning that it has to operate with other air traffic control systems around the world, such as those under development in the European Union Single European Skies Air Traffic Management Research program (SESAR), as well as Canada's system (briefly discussed in Chapter 3). These complex issues can be divided into general categories that interact and overlap in complicated ways.

SYSTEMS ENGINEERING

Systems engineering is used to manage complexity and is most valuable when applied to address the entire system throughout its life cycle. Systems engineering principles are necessary to manage the integration obstacles that must be met in order for the NextGen National Airspace System to run smoothly. These challenges, with the complexity introduced by systems maturing and also transitioning into operation on a timeline spread over more than two decades, create a strain on engineering and integration. The FAA's approach to grappling with this is explained in the FAA Systems Engineering Manual (SEM).[2] The SEM demonstrates the FAA's understanding of the principles and processes necessary to support integration and achieve lifecycle management. Unfortunately, systems engineering does not receive detailed discussion in the February 2014 Research Plan, and the plan does not even cite the SEM as a reference.

FINDING: A well-planned systems engineering approach is essential to the success of the integration and implementation of a complex multi-faceted system of systems like NextGen, and this is not adequately addressed in the February 2014 Research Plan.

---

[1] FAA, *Research Plan: Methods and Procedures to Improve Confidence in and Timeliness of Certification of New Technologies Into the National Airspace System,* Final, Office of NextGen, Washington, D.C., February 2014; reprinted in Appendix A.

[2] FAA, *Systems Engineering Manual*, Version 1.0.1, June 19, 2014, https://nasea.faa.gov/publications/main.

REQUIREMENTS DEVELOPMENT

Requirements development is an important and fundamental component of the systems engineering process. Requirements development must be devolved from and linked to the system strategic plan in terms of objectives and schedule. It is by necessity an integrated team activity involving operational components and technology developers with strong human factors influence in the case of a system such as NextGen.

The FAA Systems Engineering Manual (SEM) and Acquisition Management System (AMS) lay out clear descriptions of the importance and role of requirements analysis to include requirements development and management, and it is in the best interests of technology program managers to rigorously follow the guidelines in the AMS. The committee agrees with the governing premise laid out by the February 2014 Research Plan:[3]

> It is assumed that, from a programmatic perspective, the FAA is addressing the delivery of key technologies for the evolution of NextGen. Therefore, it is assumed that this research plan will address *the technical aspects of the approval process versus the programmatic concerns* [emphasis added].

Unfortunately, the research plan contradicts this stated assumption by focusing on the design approval process—a programmatic concern that takes place very early in the acquisition process prior to initial investment decision. The bulk of the plan's proposed outputs[4] address design approval. It is the *only* approval area of the three shown in Figure 1.1 that the plan indicated would be addressed and provides no indication of how the Acquisition Management System procedures are to be followed.

The majority of technologies needed to deliver near- and mid-term NextGen capabilities are at or approaching maturity. What the committee would expect in the plan is that it would focus on research for the approvals/certifications that are required to transition NextGen to the field and into the hands of the users.

The plan also lacks a description of a coherent concept for the research and how elements of the plan relate to the overall NextGen implementation plan. Also missing is any stated intent to conduct a shortfall analysis or identify, prioritize, and extract stakeholder needs and develop relevant requirements for the product of the research. Without defined requirements, the plan cannot state objectives, propose tasks, develop milestones, submit a budget, or define how it will measure success.

HUMAN FACTORS

Human factors may be defined as the scientific discipline concerned with the understanding of interactions among humans and other elements of a system. The field of human factors also includes the engineering profession that applies theory, principles, data, and methods with the goal of improving the human operator's contribution to system performance through improved technologies, work environments, work practices and procedures, and training. For NextGen, the key human operators include not only air traffic controllers and pilots, but also those with whom they interact, including traffic flow management, cabin crew, and airline operations. Key operators may also need to include the technicians and ground handlers who service and maintain the technologies.

The human's contribution to system performance is provided in several key ways that should be considered in certification, specifically the approval of the operational capability of the technology and its ultimate implementation by National Airspace System users. One important contribution is when humans operate and interact with technology, both as users of information and as the controllers or supervisors of automation; here, human-computer interaction, human-automation interaction, and human supervisory control of automation are important fields to consider.

Human operators also contribute to system performance in several other important ways that must also be accounted for in certification. This includes the following:

---

[3] FAA, *Research Plan,* 2014; reprinted in Appendix A, p. 4.

[4] FAA, *Research Plan,* 2014; reprinted in Appendix A, p. 5.

- *Human operators communicate and interact with each other.* Examples relevant to NextGen include Collaborative Decision Making, and trajectory negotiations conducted before and during flight that may involve traffic flow management, air traffic controllers, pilots, dispatchers, and airline operations managers. Historically, new developments directed at one operator have had side-effects in how those operators interact with others, such as changes in air traffic control required to allow more fuel-efficient descent profiles. Air and ground operators cannot be certified independently with confidence in tightly-coupled operations.
- *Human operators integrate multiple different tasks and systems.* Thus, procedures and human-integrated-systems that are developed and certified as correct independently may not function together. For example, different prototype air traffic systems have proposed different content and information for the controller's data block. Each alone may be fine for their purpose, but their combination may be confusing or overwhelming. Likewise, different air traffic procedures may each be manageable on their own, but their combination may create workload spikes or other operational blocks to effective human contributions to system performance.
- *Human operators adapt their behavior to meet goals and procedures.* Unlike technologies with fixed, deterministic functioning, human operators can vary their behavior to meet goals and procedures. However, this strength also implies that they can use the same system differently in two contexts, and thus the context of use must also be examined in certification, including operational procedures and best practices.

SAFETY AND PERFORMANCE ASSURANCE— SOFTWARE ASSURANCE AND SOFTWARE CERTIFICATION

Software assurance addresses the confidence that the software functionality performs as intended without additional side effects. The FAA defines software assurance as "the level of confidence that software is free from vulnerabilities, either intentionally designed into the software or accidentally inserted at any time during its life cycle, and that the software functions in the intended manner."[5]

The February 2014 Research Plan identifies software assurance and software certification as areas of technologies to be further investigated. The plan specifically identifies researching current and future processes of air traffic control to derive measures. However, it falls short at identifying source motivators for improving the software assurance, an assessment of the current software assurance practices, and the desired level or attributes to target in its software assurance program.

FINDING: The February 2014 Research Plan does not describe how the FAA is providing research for a robust software assurance program.

RECOMMENDATION: The FAA research plan should address software assurance issues associated with complex systems in order to ensure timeliness and confidence in the certification of new technologies into the National Airspace System.

Relevant software assurance issues include the following:

- *Software complexity.* Software has become and will continue to be prevalent in implementing existing and future functionality across the FAA enterprise, including the ground, aircraft, and associated communications. These software implementations range in size, complexity, and cohesion, highlighting the importance of addressing software assurance as a means to enhance confidence of intended and only intended behavior.
- *Staged software development and integration.* Software implementations often involve integration of modules developed over time. This is particularly true for functionality integrated across the aircraft and ground systems. The staged software development and integration may result in limited validation, analysis, and testing of the overall capability.

[5] FAA Order 1370.109.

- *Increased usage of open source and commercial off-the-shelf software.* The increased use of software from different sources introduces many challenges, including integrating software with different pedigrees with varying levels of trust. Other categories of software that may have varying pedigrees include legacy software and reusable software. The varying sources and type of software for analysis range from having detailed source code to executables.
- *Software integration into overall system.* Software is one of many components comprising future capabilities. As such, it is important that the software analysis be fully integrated in certification, verification, and validation activities. Software assurance needs to be considered in this context to improve assurance as part of integrated and timely processes and increase confidence.
- *Integrated test beds.* Test beds provide an opportunity to perform enhanced analyses targeting software assurance in a low-risk environment. Test results need to be appropriately interpreted and incorporated into the testing, verification, and validation activities.
- *Data privacy.* A system such as the current ability of a user to block FAA release of flight information for a particular aircraft has been mentioned in various FAA documents and is important as automatic dependent surveillance-broadcast (ADS-B) is implemented.
- *Resilience to cyberattack.* Software assurance activities need to be included as a dimension of the overall cybersecurity strategy and implementation targeting continued system operations during cyberattack. Systematic approaches to cybersecurity are a fundamental part of effective and improved future technologies and capabilities. Cybersecurity spans the development and operational activities including failure/anomaly analysis and integrity of the software development tools and environment.

FINDING: Cybersecurity is a critical component of the National Airspace System that needs to be addressed early, continually, and comprehensively across the systems, segments, and procedures.

There are increasing indicators of cyberattacks targeting and exploiting various programs, technologies, and stakeholders. Consequently, cybersecurity should be addressed as an integral part of delivering services to ensure that the threat environment be considered as part of the architecture, design, development, systems engineering, and operational activities.

RECOMMENDATION: The FAA research plan should address cybersecurity as an integral part of the National Airspace System.

SAFETY AND PERFORMANCE ASSURANCE—VERIFICATION, VALIDATION, AND TESTING

The processes of validation and verification generally are applied to software, but in the context of certification and implementation for NextGen, they are integral parts of software, processes, and hardware development, procurement, and application. To be effective, these have to be conducted across the entire FAA organization.

Verification is generally defined as the detailed process of review and evaluation of products through the completion of the development phase to assure that it functions properly and meets the specified requirements for which it was intended. Validation examines and determines if the product satisfies specified business or user requirements. The objective is to determine if the product meets the user's needs; it may also address whether the specifications were correct in the first place. Together, verification and validation of the overall system demonstrate that the product meets the need when placed in the user environment.

The management of the validation and verification processes extends to the procurement phase, which may include multiple contractors with differing verification and validation approaches. The management also has to deal with the rate of advancements in technology, specifically software and the changing operational needs and the management complexities within the organization. These factors highlight the need for a strong policy on software development, deployment, and maintenance.

In addition to policy needs, the committee notes the importance of maintaining a strong and capable staff to address all phases of validation and verification. This is particularly true in the area of software development and procurement. The rapid evolution of technology in this area points to the need for training in addition to a capable team of staff and management personnel.

The implementation of new technology involves user equipage, and the En Route Automation Modernization system (ERAM) is a good example. Implementation took years, and operational testing was done incrementally in a series of "builds": keyboard entries, hand offs, flight plan amendments, altitude changes, etc. Each step along the way had to be tested to ensure correct interfacing with legacy systems while adhering to a requirement to continue services without interruption.

Introducing a new operational capability that requires user equipage is much more complex than implementing procedural changes and requires user investment and time. The benefits do not accrue until users equip, and users will not equip unless they are confident in the ability of the FAA to deliver, not only on new technology, but also new operational benefits that have an economic payoff on a reliable schedule.

Whereas avionics improvements happen at a relatively fast pace, it takes much longer to verify, validate, and test their operational performance. The FAA was charged to develop a plan to increase user confidence that the FAA would deliver promised operational capabilities with an economic payoff on a stable schedule. But the relative lack of attention to the challenges associated with verification, validation, and testing makes it difficult to see how users could gain confidence without knowing how this will be done.

FINDING: Verification and validation is a very complex process. While much research has been done in this area, it is cost and time intensive. Great reductions in certification time could be achieved if deliberate and focused research resources were devoted to this area.

RECOMMENDATION: The FAA research plan should include as a significant priority the improvement in the use of verification and validation of the overall system. The FAA research plan should demonstrate how the FAA is building on the significant research on verification and validation being done by NASA and other government research laboratories, academia, and international research groups.

TRAINING

The training for and the maintenance of proficiency of operational personnel in the proper use of systems is essential to the successful integration of new technologies and procedures into the National Airspace System. The processes for FAA approval (certification) of training programs are often time consuming and generally complex. A regulation is usually issued mandating the training. This step is followed by the creation of an advisory circular offering an example of a compliant training program. In turn, that step is followed by written "guidance" to the FAA local certificate management offices (CMOs) delineating standards for acceptance of an individual airline training program. Using these documents, airlines then develop individual training programs for submission to their CMO for approval. Typically, the CMO submits the draft training program to the FAA in Washington, D.C., for the concluding review before a final letter of approval is issued. A similar process is followed for air traffic controllers.

The February 2014 Research Plan does not specifically address training issues. These issues are only discussed vaguely, and the plan only addresses research on the approval process for the implementation of new ground-based technology, processes, and procedures, as opposed to a fully integrated system and its components. The plan could be more robust and useful if it addressed the steps required in the development and approval of training programs using process mapping and redundancy elimination tools. The number of times stakeholder groups are required to comment in the approval process is an issue that requires close attention. Although training is not inherently a "research" issue, NextGen creates challenges for pilots and ground controllers. It is possible for the FAA to conduct research on better training methods, and there may also be ways to streamline and expedite the certification and training processes as an integral part of the National Airspace System.

OPERATIONAL TEST AND EVALUATION

The FAA has a long-standing operational test and evaluation capability that supports its development and deployment of equipment and procedures. As both ground and airborne equipment is developed and ultimately certified, the operational evaluation plan is used to ensure that requirements are accurate, beneficial, and able to be implemented.

One of the most critical aspects is evaluating the new capability and its smooth integration into the existing technology baseline. Safety and performance requirements for the air traffic system are extremely high. Overall, the operational evaluation program is designed to ensure there is no degradation in performance as new capabilities are integrated into the National Airspace System. To the extent feasible, the intent is to validate that new capabilities add to system safety and performance. The FAA's operational evaluations take a multi-disciplinary approach and incorporate participants from multiple offices and lines of business. External vendors and the frontline workforce representatives are also actively involved. The teams are established to operate with independence, to ensure a lack of bias, and to focus on safety outcomes.

The February 2014 Research Plan mentions how new technologies intended to benefit the users will need to be evaluated and that specific outputs for timeliness and confidence in the areas of verification and validation, testing, and operational evaluation will be part of the program. The project schedule, milestones, and budgeting table[6] lists a gap analysis on the long-term needs, and an analysis of new processes, procedures, and technologies. However, the committee could not determine if these activities, as briefly described, would adequately address the requirement for detailed and continuing operational evaluation. Operational test and evaluation can possibly be streamlined to take fuller advantage of the new and emerging technologies and capabilities.

CONCLUDING REMARKS

Despite the complexity of the task, the committee notes that the FAA has demonstrated the ability to do successful integrated projects as part of NextGen, and these have already had substantial impacts on improving operations within the National Airspace System. Several examples of these are addressed in Chapter 3, along with some relevant lessons from other organizations.

[6] FAA, *Research Plan,* 2014; reprinted in Appendix A, p. 7-9.

3

Learning from Other Projects and Other Organizations

In recent years, the FAA and the aviation industry have successfully demonstrated the value of FAA research to improving the safety, efficiency, and quality of aviation in the United States. This chapter summarizes three such programs that went from research to operations: the Alaska Capstone Program, the Gulf of Mexico Helicopter initiative, and the Greener Skies Over Seattle initiative.

ALASKA CAPSTONE PROGRAM

The Alaska Capstone Program was primarily a safety improvement program in Alaska. Alaska poses significant environmental and operational challenges to aviation, and the state is heavily reliant on aviation to connect communities across vast distances. The accident rate in Alaska is much higher than in the lower 48 states, which prompted the FAA to initiate the Capstone program. The specific areas of concentration the program recommended to improve on were to reduce the accident rate due to navigation errors, such as controlled flight into terrain, and to reduce the rate of mid-air collisions.

A fleet of small commercial aircraft initially evaluated safety benefits of technologies during day-to-day operations in Alaska. The aircraft were fitted with instrument flight rules (IFR) capable systems. These included Global Positioning System (GPS) receivers; a universal access transceiver (UAT) data-link system that enabled automatic dependent surveillance-broadcast (ADS-B) and flight information service (FIS), including real-time weather. Also included was a multifunction display depicting terrain, other ADS-B aircraft, and weather graphics and text data. Approximately 200 aircraft used for commuter, charter, and mail flights in southwest Alaska were equipped.

The goal was to identify an affordable means to reduce controlled flight into terrain, providing pilots with an enhanced means to see nearby traffic and receive current weather in the cockpit. The Capstone program provided training for pilots, operators, safety inspectors, air traffic control specialists, and technicians. To enable air traffic services (ATS) to use ADS-B in the Bethel non-radar environment, Anchorage Air Route Traffic Control Center's equipment was modified so that the air traffic controller's display showed both radar and ADS-B together. This system operated from 2000 to 2006 when it was merged with the National ADS-B program. A safety analysis done by the University of Alaska showed a reduction of 47 percent in the accident rate of equipped aircraft between 2000 and 2004.[1]

[1] FAA, "Alaska Controllers Use Next Generation Air Transportation Technology to Improve Safety," Press Release, June 24, 2010, https://www.faa.gov/news/press_releases/news_story.cfm?newsId=11540.

GULF OF MEXICO HELICOPTER INITIATIVE

Before December 2009, the 600,000 square miles of the Gulf of Mexico had little surveillance and was characterized by large separation standards and difficult flying conditions, particularly in Instrument Meteorological Conditions (IMC). ADS-B, first operational in 2010, has been helping helicopters reach offshore oil rigs in poor weather conditions that would have previously left supplies and rig workers back at the airport unable to fly.

In the past, because of insufficient radar coverage over the ocean, when bad weather obstructed visibility in the Gulf and pilots had to navigate using onboard instruments rather than visually avoiding other helicopters, the FAA imposed severe restrictions on helicopter operations. The FAA either grounded the aircraft or required that each helicopter stay in its own 20-by-20-mile airspace grid. This created problems because air traffic in the Gulf is nearly as busy as the heavily traveled East Coast corridor, with some 5,000 to 9,000 helicopter offshore platform operations every day.

In December 2009, the FAA debuted the ADS-B in the Gulf. In a joint agreement with the FAA, Helicopter Association International, helicopter operators, and oil platform companies, the federal government installed ADS-B transmitters, along with weather observation and communications equipment, on 12 offshore platforms; and operators equipped their helicopters with ADS-B avionics.

Similar to radar, ADS-B pinpoints the location of helicopters and aircraft. FAA air traffic controllers now provide the same surveillance and air-to-ground communications to helicopters in the Gulf as they do for aircraft flying over land. Helicopters can also use the same standard 5-mile separation distance when flying over the Gulf. This reduction to 5 nautical miles from a 20-by-20-mile block allows direct routing clearances for ADS-B-equipped helicopters. The FAA estimates more than 300,000 nautical miles in flight savings from December 2009 to February 2014.[2] An individual helicopter operator reports an increase in its average annual IFR flight hours from about 1,500 to about 20,000 with all its helicopters equipped with ADS-B. The users cite schedule dependability as the biggest benefit for their operations.

GREENER SKIES OVER SEATTLE[3]

On June 11, 2012, a commercial passenger flight landed at Seattle-Tacoma International Airport using a satellite-based navigation arrival procedure that may significantly reduce aircraft exhaust emissions and fuel burn, contributing to better air quality in the United States and eventually around the world. This was done as part of the Greener Skies Over Seattle initiative, a collaborative project between the FAA, airlines, the Port of Seattle, and Boeing Corporation. This initiative involves a combination of new technology, procedures, flight regulations, and training. The FAA will add 27 new procedures, expanding the use of Optimized Profile Descents, where the airplane essentially glides in idle to the runway threshold; Area Navigation (RNAV) arrivals, which are GPS-guided arrivals; and Required Navigation Performance (RNP) approaches, which take RNAV to an additional level of precision. These procedures were planned to be available to any properly equipped aircraft by spring 2015.

Greener Skies aims to prove that satellite-based navigation approaches can be flown using the same separation standards as the current procedures using ground-based instrument landing systems. The trials seek to determine that a curved RNP approach to one runway is so precise and predictable that when it is flown next to another aircraft that is approaching a parallel runway, it merits the same separation standard as two straight-in parallel approaches. Although the project is still under way, it demonstrates how the introduction of new air and ground systems and procedures can have beneficial effects on air transportation.

[2] FAA, *NextGen Implementation Plan*, August 2014, https://www.faa.gov/nextgen/library/media/NextGen_Implementation_Plan_2014.pdf.

[3] FAA, *NextGen Implementation Plan*, 2014.

CHARACTERISTICS OF NEXTGEN SUCCESSES

All three successful NextGen programs share the following characteristics:

- Stakeholders were involved from the beginning of the program.
- The project plan had the commitment of each stakeholder.
- Frequent progress report meetings were held with project reviews of all actions as a group.
- Stakeholders stayed committed throughout the project.
- Projects were directed at specific sites or specific areas. Scaling it up for the entire United States is not automatic (i.e., "do the same everywhere") and presents major challenges of integration with the existing ATC systems, procedures, mixed equipage, and the diversity of users. Stakeholders' communities are much larger and very diverse, hence, creating and *holding* consensus for the duration of any project is a major challenge.
- The stakeholders in these programs received easily observable benefits. It is not at all clear that this is the case for NextGen.

The Alaska Capstone Program, the Gulf of Mexico Helicopter initiative, and the Greener Skies Over Seattle initiative demonstrate some of the potential benefits of the components of NextGen, as well as some of the attributes required for the successful implementation of a program. They are included here because they can help guide the development of future FAA research programs and indicate what aspects are necessary for a successful program. However, these projects were limited in scope and do not demonstrate the scale and complexity that will be needed for FAA to be successful with NextGen, or its larger research and development and deployment mission.

OTHER AGENCIES AND ORGANIZATIONS

During the course of this study, the committee sought to determine how other agencies and organizations engage in certification of new technologies. The committee was briefed by managers from the U.S. Air Force Flight Standards Agency and the Air Force Operations Systems Requirement Group, and from NavCanada.[4]

The Air Force Flight Standards Agency briefing showed an effective system of linking requirements to the certification plans for weapons systems. This very active certification program and the Air Force's experience may provide the FAA with some lessons learned. However, this weapon system certification program does not appear to be directly relevant to the FAA.

The Air Force reported working closely and in cooperation with the FAA on air traffic control requirements and accepting FAA direction on implementing new systems. An Air Force representative indicated that the Air Force is required to meet other nations' air traffic control requirements that, in some cases, have advanced beyond the current FAA concepts. These requirements have caused operational compromises for the Air Force in some cases.

A NavCanada briefing to the committee highlighted their methods of implementing new technologies and operations. While Canada has the second largest airspace service provider in the world geographically, it is still smaller than the U.S. National Airspace System, and some of Canada's initiatives have been addressed with implementations specific to certain types of operation or facilities. Thus, NavCanada's methods may not scale to larger FAA NextGen initiatives. However, notwithstanding that limitation, the NavCanada approach was notable for its emphasis on simultaneously developing new operations and the technologies that would enable these operations. The NavCanada approach uses embedded teams that include stakeholders, and it has demonstrated a track record over the past few years of repeated, incremental change. In some areas, NavCanada systems surpassed FAA systems. Further, the NavCanada managers described consciously deciding to scope each incremental change to be small enough to remain tractable, with each change fitting into a larger strategic roadmap of changes planned in the near future.

[4] NavCanada is the company that owns and operates Canada's Air Navigation System.

A number of elements have led to NavCanada's success:

- The certification organization is independent of the program management offices.
- The process is well-defined and data-driven.
- Endorsed technical and operational experts are involved throughout the process, including the developmental phase, and it is clear what operations are to be enabled.
- It is known in advance when new certification requests will be expected.
- The process is flexible enough to provide incremental improvements as determined during the process.
- All stakeholders are aware and involved at each stage as needed.
- Accountability at each stage is well understood by all.

While acknowledging that NavCanada is different and faces different issues, the committee concluded that the FAA could learn valuable lessons by studying how NavCanada operates and has achieved success in some instances.

Finally, although the committee was not specifically briefed by NASA, its members are aware of the extensive research conducted by NASA on many aspects of air traffic control in general and NextGen in particular. FAA and NASA work closely together on this research. However, the February 2014 Research Plan does not refer to this important relationship. The committee considers this to be a deficiency and believes that any future research plan will not be complete unless it references relevant NASA research as well as that of other agencies and organizations that develop air traffic control technologies or are involved in certifying new technologies.

FINDING: The February 2014 Research Plan does not refer to research conducted by NASA.

RECOMMENDATION: The FAA research plan should benchmark the best practices regarding certification of other organizations that can contribute to the timely implementation of NextGen technologies and coordinate its research with other relevant organizations, particularly NASA.

4

Effective Communication Between the FAA and Congress

During the course of this study, the committee conducted discussions with the FAA staff directly involved in the generation of the February 2014 Research Plan,[1] congressional staff who authored the original legislation that called for the research plan, foreign and other U.S. government agency personnel involved in similar activities, and aviation industry representatives. The committee also heard from other members of FAA management. From these discussions, the committee noted that significant inconsistencies in expectations exist in relation to what will be conducted and who benefits from NextGen initiatives, when the benefits will be achieved, and who controls the implementation.

Since the FAA began initiating upgrades to the National Airspace System over a decade ago, there have been substantial changes in the underlying conditions that initially justified NextGen. The overcrowding of the airspace that resulted in air travel gridlock in the previous decade has significantly decreased, and safety concerns have been alleviated with a continuous statistical improvement in U.S. aviation safety as the number of flights in the National Airspace System increased.

Based on the committee's discussions and the observation that the original driving forces that helped initiate NextGen have weakened in the current environment, the committee concluded that measurable performance objectives for NextGen will likely change. The committee also concluded that there is a need for champions to lead specific elements of NextGen's implementation. Emerging disruptive technologies or new systems, such as the introduction of Unmanned Aerial Vehicles (UAVs)[2] into the National Airspace System, may also provide a significant requirement to cause accelerating the implementation of NextGen capabilities. These will have broad system-wide applications.

In addition, based on past experience, the committee is aware that the FAA has a dedicated and capable group of practitioners who can implement the complex systems required for a fully capable NextGen. However, cultural barriers to cross-departmental and agency collaboration limit full integration of the multiple projects in a timely way.

The FAA informed the committee that it has plans for aviation research and is committed to enhancing the certification process that leads to new capabilities and improved system safety and performance. The FAA is primarily an operating agency providing air traffic control services, and its culture reflects the priorities associated with operating an around-the-clock air traffic system that is the largest, most sophisticated, safest, and most

[1] FAA, *Research Plan: Methods and Procedures to Improve Confidence in and Timeliness of Certification of New Technologies Into the National Airspace System,* Final, Office of NextGen, Washington, D.C., February 2014; reprinted in Appendix A.

[2] Also often referred to as Uninhabited Aerial Systems (UAS).

diverse in the world. The core attribute of the system is its extremely high safety level, and the key mission FAA has is to achieve annual improvements in the safety level.

Adding new capabilities can introduce risk into the system and, therefore, new capabilities are subjected to thorough analysis and testing prior to implementation. The required safety and performance levels for these capabilities means that progress will be deliberate, incremental, and evolutionary. However, in the end, NextGen will be transformational. In addition, the supporting research activities as well as the processes associated with planning and identifying operational benefits will also be incremental as the FAA pursues the essential engagement with dozens of external stakeholders.

The modernization of the National Airspace System is a highly integrated process and in a program that requires the incorporation of needs both internal and external to FAA control and influence. For an operational change and benefit to be deployed and implemented, the agency has to coordinate thousands of people and multiple levels of approval, frequently over many years and changing environmental conditions.

A good recent example of a relevant FAA initiative is the *Navigation Procedures Initial Implementation Plan (NAV Lean)* program.[3] This targeted and pragmatic initiative by FAA leadership has thoroughly reviewed its certification processes and identified useful steps to streamline and improve its ongoing evolution. The benefits are a certification process that is more responsive in less time and at lower cost to applicants, while also being rigorous, thorough, and continuing to ensure public safety outcomes.

During the course of this study, including interviews with many government employees and industry stakeholders, it became clear that the February 2014 Research Plan was clearly a missed opportunity for FAA to reassure Congress and stakeholders about much of the excellent progress that NextGen has already achieved, as well as the various efforts currently under way to demonstrate confidence in the timeliness of certification of new technologies into the National Airspace System.

The committee also concluded that there is widespread lack of a knowledgeable understanding of all of the complex aspects of this problem. The committee hopes that this report can serve as a catalyst for more effective communication and collaboration between Congress and the FAA.

The FAA is a vital 24-hours-7-days-a-week operating agency providing critical public transportation safety services. It does not have a culture oriented toward skillfully communicating its programs or successes to Congress or other stakeholders. Yet Congress performs an essential service by conducting oversight and providing funding for the FAA that depends on knowledge of the FAA's management efficiency, productivity, and clear strategic implementation.

The February 2014 Research Plan does not provide a robust or insightful outline of the FAA's direction or integrated programs. In fact, the plan seemed overly cautious and excessively edited, thus limiting its usefulness. The committee concluded that Congress was actually seeking a broad answer that went beyond simply the FAA's plans for "research," but the FAA responded in a narrow fashion, after giving the response a low priority. The congressional request was well intentioned but insufficiently clear about what Congress actually was seeking. The FAA's response (the February 2014 Research Plan) was too limited to actually be helpful to Congress, the FAA, and its stakeholders and clarify the current situation and efforts to make progress.

The only way to break the cycle of miscommunication that this report reflects is for the FAA to prioritize its responses to Congress and actively communicate its actions and integration strategies. The FAA is streamlining processes to speed the procedures approval process and has involved stakeholders, like air traffic controllers, more actively in its integrating process. Overall, the FAA is following the priorities established by its key advisory group of stakeholders in the NextGen Advisory Committee.

Numerous studies and reports have recognized the extremely complex and challenging nature of FAA's certification and system modernization tasks, most recently in an October 2014 MITRE study.[4] In addition, the FAA tasked the School of Systems and Enterprises of the Stevens Institute to study lessons learned from comparable

[3] FAA, *Navigation Procedures Initial Implementation Plan (NAV Lean)*, June 1, 2011, http://www.faa.gov/nextgen/media/SIGNED%20Initial%20NavLean%20Implementation%20Plan%201%20June%202011.pdf.

[4] MITRE, *NextGen Independent Assessment and Recommendations*, MITRE Project No. 0214DL01-IF, Center for Advanced Aviation System Development, McLean, Va., October 2014.

large-scale, complex, systems-integration projects.[5] Also, the Stevens Institute developed a competency model for acquiring large, highly complex systems of systems and prepared a report that was published by the International Council on Systems Engineering.[6]

FAA's certification activities that are supported by associated research are necessarily conservative and methodical and occur at a measured pace. Congressional expectations for clarity, efficiency, and annual progress are reasonable. However, neither the legislative language that called for a research plan nor the plan that the FAA eventually produced were clear and unambiguous. There is also the reality that a research plan is only one part of a much broader series of actions and activities that ultimately determine the implementation of new technologies. The FAA is not and cannot be a command and control authority driving the aligned investments required by a broad and diverse industry with corporations that have their own business cases to evaluate and pursue.

The committee believes that FAA's future responses to Congress should not be narrowly constructed, one-time communications that lack context. The February 2014 Research Plan was too narrow in scope to provide an adequate response to what Congress was seeking. It would enhance FAA's credibility and effectiveness if the FAA ensured that the readers of its communications were presented with a wider context of its key activities and investments as well as guidance on how to interpret them.

---

[5] R. Turner, D. Verma, and W. Weitekamp, *The Next Generation Air Transportation System (NextGen)*, School of Systems and Enterprises, Stevens Institute of Technology, Hoboken, N.J., August 31, 2009.

[6] J.R. Armstrong, D. Henry, K. Kepchar, and A. Pyster, Competencies required for successful acquisition of large, highly complex systems of systems, *INCOSE International Symposium* 21(1):629-647, 2011.

Appendixes

A

FAA Research Plan

The Federal Aviation Administration (FAA) report *Research Plan: Methods and Procedures to Improve Confidence in and Timeliness of Certification of New Technologies Into the National Airspace System* (Final, Washington, D.C., February 2014) is reprinted here without its contents page or list of acronyms.

Research Plan

Methods and Procedures to Improve Confidence in and Timeliness of Certification of New Technologies Into the National Airspace System

FINAL

February 2014

Change History

Version	Date	Change
1-00	5/24/2012	Baseline Version
2-00	8/20/2012	Final Draft
3-00	10/3/2012	Update to Final Draft
4-00	11/28/2012	Final
5-00	1/3/2013	Final-Cleared
6-00	4/9/2013	Final-Adjudicated
7-00	1/27/2014	Final-Revised

Prepared by:

Office of NextGen
William J. Hughes Technical Center
Aviation Research Division

EXECUTIVE SUMMARY

On February 14, 2012, the President signed the FAA Modernization and Reform Act of 2012 (the Act), [1] Section 905 of the Act, entitled "Research on Design for Certification," instructs the Federal Aviation Administration (FAA) to "conduct research on methods and procedures to improve both confidence in and the timeliness of certification of new technologies for their introduction into the national airspace system." The Act also instructs the FAA to develop a research plan that contains objectives, proposed tasks, milestones, and a 5-year budgetary profile, and to arrange an independent review of this plan by the National Research Council.

The scope of this research plan is defined in part by 49 United States Code, Section 44505, entitled "Systems, procedures, facilities, and devices." In preparing this report, we have considered certification processes for the two basic categories of technologies that are central to the Next Generation Air Transportation System (NextGen): (1) those associated with aircraft; and (2) those associated with ground-based systems and air traffic control (ATC).

The procedures we follow to certify the manufacture and installation of aircraft equipage (i.e., avionics that enable NextGen capabilities) are critical to the success of NextGen and the aviation community's confidence in it. The FAA has completed a number of initiatives focused on the manner and procedures for approving new technologies in aircraft. Most recently, the FAA convened industry experts in the Aircraft Certification Process Review and Reform Aviation Rulemaking Committee (ARC), as required by Section 312 of the FAA Modernization and Reform Act of 2012. The FAA is evaluating the recommendations from this committee and determining implementation plans. Therefore, at this time, the FAA has not specified additional research requirements necessary to improve the effectiveness or timeliness of certification procedures for new aircraft technologies. However, we have identified several key technical areas within the ground-based system and ATC design approval processes that would benefit from further research to determine how their effectiveness and accuracy can be improved.

New ground-based systems and ATC technologies are managed throughout their lifecycle through the FAA Acquisition Management System (AMS) [2], a process designed to increase the quality, reduce the time, manage the risk, and minimize the cost of delivering safe and secure services to the aviation community and flying public. The FAA has been using the AMS process for more than 16 years, and after a recent system-wide evaluation, we have supplemented it with Air Traffic Organization process improvement efforts and Safety Management Systems [3]. AMS areas on the critical path for programs that we believe would benefit from additional research assistance are software assurance, software certification, computer human interfaces, technical interfaces, hand-off interfaces, and verification and validation.

The FAA believes that we provide sufficient oversight to ensure that new technologies, from a programmatic perspective, are efficiently incorporated within the National Airspace System (NAS) and that associated benefits are achieved for the aviation industry in a timely and accurate manner. These oversight vehicles include the formation of the Office of NextGen (ANG), which reports directly to the Deputy Administrator of the FAA; the focus on NextGen transformational programs (i.e., Automatic Dependent Surveillance-Broadcast) that provide core capabilities in achieving the NextGen vision; the incorporation of portfolio capture teams that shepherd related capabilities through their lifecycle; and portfolio leads, who ensure appropriate integration of systems and capabilities.

This plan fully addresses the scope, tasking, and funds required to conduct the identified research. As a result of this planning effort, a 5-year research budget of $4.6M is required to execute this research plan, make appropriate enhancements to the approval process, and develop associated information technology. This research plan is not currently included or funded in the National Aeronautics Research Plan (NARP), partially because it needs to go through an independent review and adjudication process with the National Research Council, as described below.

1. BACKGROUND AND SCOPE

The Next Generation Air Transportation System, or NextGen, is a set of initiatives that will make our national airspace system (NAS) safer and more efficient while curbing aviation's environmental impact.

NextGen's suite of tools includes better traffic management and improved procedures that will help us reduce delays and save fuel. Satellite-based surveillance provides more precision in tracking aircraft, and gives pilots the ability to see other aircraft around them just as air traffic controllers do. Advanced digital communications between the ground and the flight deck reduces opportunities for error, and system-wide information management gives the right information to the right people at the right time.

The FAA is challenged with implementing a complex set of NextGen improvements into a NAS that operates 24 hours a day, every day. One of our challenges is earning the confidence required for operators to equip their aircraft to take advantage of NextGen capabilities as they become available.

On February 14, 2012, the President signed the FAA Modernization and Reform Act of 2012. [1] Section 905 of the Act, entitled "Research on Design for Certification" instructs the Federal Aviation Administration (FAA) to "conduct research on methods and procedures to improve both confidence in and the timeliness of certification of new technologies for their introduction into the national airspace system." The Act also instructs the FAA to develop a research plan that contains objectives, proposed tasks, milestones, and a 5-year budgetary profile, and to arrange an independent review of this plan with the National Research Council.

The scope of the research plan is dictated in part by the language in Section 905 that amends 49 United States Code (USC) Section 44505, "Systems, procedures, facilities, and devices." The primary focus of Section 44505 is to meet the needs for safe and efficient navigation and traffic control of civil and military aviation. More specifically, this section addresses research on human factors and simulation models, and developing and maintaining a safe and efficient system.

The FAA has convened industry experts in the Aircraft Certification Process Review and Reform Aviation Rulemaking Committee (ARC), as required by Section 312 of the FAA Modernization and Reform Act of 2012, and the Office of Aviation Safety (AVS) has developed comprehensive plans for the implementation of aircraft certification technologies, including an Aviation Safety Work Plan for NextGen 2012 [4] that addresses the aircraft certification activities for new technologies. Therefore, certification activities specifically related to aircraft, aircraft engines, propeller, and appliances (as described under 49 USC, Section 44504) are not included in the scope of this research plan. This plan does address areas that require close coordination (i.e., software assurance, air-ground communications/interfaces, and controller training).

The AMS [2] process governs the acquisition of new technologies for ground-based systems. Continuing improvements to the AMS and Safety Management System processes will ensure that new technologies are incorporated within the NAS in a timely and safe manner. In addition, other process improvement efforts provide coordination of program portfolios that ensure the benefits of NextGen technologies are delivered in a consistent and timely manner within the NAS.

Incorporation of new technologies is intended to enhance the efficiency of the NAS (i.e., reduced separation, preferred routing, and fuel efficiency) or to increase the productivity and safety of travelers, air traffic controllers, and pilots (i.e., increased throughput, reduced operational errors, reduced pilot deviations, and shared situational awareness). Whenever new technologies are incorporated for efficiency, safety, or enhanced productivity, there is typically an impact on three critical areas:

1. New technology integration almost always includes a complex software development effort along with detailed internal and external interfaces that allow the technology to interoperate with other NAS components and external NAS stakeholders,

2. New technologies intended to benefit the users typically entail enhancements to the human interfaces (e.g., air traffic controllers and facility management personnel) involved in ATC operations, and

3. Integration of new technologies within the NAS typically affects other key areas that include, but are not limited to, key elements of the AMS process including requirements, verification and validation, and training.

An overview of the approval process is depicted in Figure 1 below. The research effort will evaluate the entire approval process with a focus on how the introduction of new technologies is handled during specific steps of the process. The research focus areas are shown capitalized within the System Design Approval, and this effort will not address the other two categories.

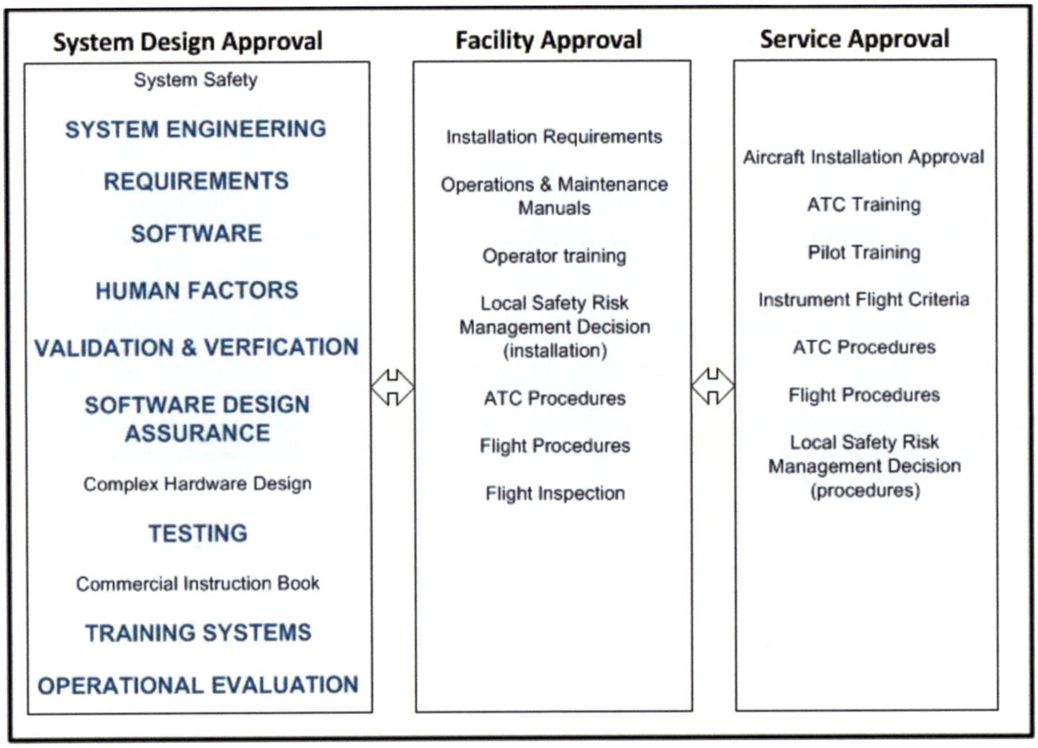

Figure 1: – NAS Ground-System Approval Process Overview

Additionally, the research will examine identification, development, and implementation of measures that would provide FAA managers and oversight personnel with objective insight into the status of the ground-based system design approval process.

2. ASSUMPTIONS

It is assumed that this research will only address the approval of new technologies being integrated into the NAS and will not specifically address aircraft or aircraft equipage certification activities. Aircraft equipage is covered under Section 215 of the FAA Modernization and Reform Act of 2012. Likewise, aircraft certification procedures are addressed under Section 312 of the FAA Modernization and Reform Act of 2012.

It is assumed that when the term "approval process" is used throughout this document, it is referring to the focus areas identified in Figure 1 of this report.

For the purposes of this research plan, the term "timeliness" refers to the overall time to define, develop, test, and deploy a design for ground or air-ground systems within the NAS.

For purposes of this research plan, the term "confidence" refers to the expectation that a system will perform within the specified reliability, maintainability, and availability as described in the system design. Likewise, the term confidence also means that a system design is fully and completely verified and validated.

For purposes of this research plan, the term "certification" refers to the overall "approval process" for the design of a ground or air-ground system within the NAS in accordance with the AMS. In addition, the term certification refers to the approved design of critical software components (i.e. software certification levels) and ground based equipment (i.e. ILS certification).

It is assumed that, from a programmatic perspective, the FAA is addressing the delivery of key technologies for the evolution of NextGen. Therefore, it is assumed that this research plan will address the technical aspects of the approval process versus the programmatic concerns.

3. RESEARCH OBJECTIVES, OUTPUTS, AND OUTCOMES

The major objective of this research is to identify methodologies, procedures, and information technologies that can increase the efficiency (i.e., timeliness) and effectiveness (i.e., confidence) for integrating new technologies within NAS ground- or air/ground-based systems (integrated NAS systems). As new methodologies and procedures are explored, there is an opportunity for both the FAA and the aviation community to achieve process improvements and minimize costs.

By researching the current and planned processes of ATC software assurance, software certification, human factors, interfaces, and training, it will be possible to create some objective measures surrounding the current and future processes. Properly analyzing this information could give FAA managers better insight into the efficiency and effectiveness of ongoing procedure approval efforts. A more forward-looking objective would be to continually monitor the process, which will allow the Air Traffic Organization (ATO) to take a proactive approach for overseeing the process and allow for more transparency into the process. Overall, all stakeholders can better understand the impacts and benefits of new technologies before final implementation.

Outcomes (i.e., timeliness and effectiveness) are not specifically measurable. More specific outputs for timeliness and confidence (i.e. measurable outputs) will be developed and delivered for each of the following AMS [2] research focus areas:

- Software assurance,
- Software certification,
- Requirements management and traceability,
- Human technologies and computer human interaction,
- Verification and Validation, Testing, and Operational Evaluation,
- Controller Training Systems,

- Technical Interfaces, and
- Information Security.

For each of the above research focus areas, the following outcomes and associated outputs will be developed:

Outcome—Timeliness

- Output—Measure time and collect data for the current design approval process
- Output—Measure time and collect data for the individual process elements of the design approval process
- Output—Collaboratively identify and recommend modifications to the current design approval process to increase timeliness
- Output—Prototype modifications to current approval process and develop/modify tools (e.g., human interaction, interface simulation, software analyses, modeling, and information security)
- Output—Apply prototype modifications to several ongoing approval process efforts (pilot studies)
- Output—Measure time and collect data on the new approval process
- Output—Produce ongoing measures and trends of the approval process
- Output—Integrate measures and trends into ATO oversight and decision-making process
- Output – Report the research results

Outcome—Confidence

- Output—Develop analysis capabilities that support air traffic personnel in evaluating information related to the design approval process
- Output—Develop information-sharing technologies that support the FAA and industry involvement earlier in the design approval process

4. RESEARCH TASKING

As instructed in the Act, the FAA will ask that the National Research Council perform an independent review of this plan and work cooperatively with the FAA to modify the plan in accordance with the review.

The research will undertake a systematic approach to achieve the required outcomes and objectives.

1. Perform background analyses and data collection of current processes and technologies and plan for addressing new technologies and planned NextGen enhancements.
2. Conduct interviews with current FAA and industry personnel who frequently perform the design approval process (e.g., software assurance, software certification, human factor assessment, training development, and information security).
3. Identify measures and potential enhancements to the processes, procedures, and information technologies surrounding the design approval process.
4. Perform a gap analysis on the long-term needs and currently funded initiatives.
5. Design modifications to current design approval processes to help increase the timeliness and effectiveness of the process.
6. Design and develop information technology and human performance analysis prototypes to support the design changes and identified measures. Develop applicable tools and facilities to support research focus areas (e.g., software assurance, software certification, requirements, verification and validation requirements).
7. Analyze new processes, procedures, and technologies (benefits analysis).
8. Implement new processes, procedures, and technologies (via pilot studies of new techniques and technologies) that will result in enhanced design approval processes.
9. Monitor and collect data on new processes, procedures, and technologies.
10. Validate measures and results of new processes, procedures, and technologies.
11. Produce a final report with results, findings, and suggestions for recommended modifications of processes, procedures, and applicable technologies that can be integrated into the ATO workforce.
12. Integrate new modifications within the ATO.

5. CHALLENGES

Several known challenges are affiliated with this type of research effort.

While the agency embraces positive change, it can take large organizations time to assimilate it throughout the operation. Careful steps must be taken to educate the workforce participating in the pilot projects so there is an awareness that the research effort is of benefit to them. Research results will allow for a more efficient and timely process so procedures can be developed faster without degrading the effectiveness of the process.

FAA also needs to invest in efforts with a positive cost-benefit ratio.

6. PROJECT SCHEDULE, MILESTONES, AND BUDGETING

This section describes the key dates anticipated from the research effort. Key dates for the project should be integrated with all stakeholders for the research project.

Task	Deliverable	Date	Funding
Develop Research Plan	Draft Research Plan	Aug 2012	$ 0K
Release Research Plan	Final Research Plan	Dec 2012	$ 0K
Perform background analysis and data collection of current processes and plan for addressing new technologies and planned NextGen enhancements.	Analysis Report	Sep 2013	$200K
Have the National Research Council perform a formal review of the Research Plan and coordinate required tasking and outputs of research activities.	Research Plan Update	Dec 2013	$500K
Conduct interviews of current FAA and industry personnel who frequently perform the approval process (e.g., software assurance, software certification, human factor assessment, training development, and information security).	Research Report—Design Approval Process, Tools and Facility Requirements	Mar 2014	$200K
Identify measures and potential enhancements to the processes, procedures, and technologies surrounding the design approval process.	Research Report—Description of measures for monitoring design approval process	May 2014	$200K
Perform a gap analysis on the long-term needs and currently funded initiatives.	Research Report—Gap analysis of current design approval process	Jul 2014	$200K
Design modifications to the current design approval processes to help increase the timeliness and effectiveness of the process.	Research Report—Suggested enhancements to design approval process	Jul 2015	$500K
Design and develop information technology and human performance analysis prototypes to support the design changes and identified measures.	Prototype Application—Design Certification Information Systems	Jul 2016	$1.5M
Analyze new processes, procedures, and technologies (benefits analysis).	Research Report—Benefits analysis of modifications to the design approval process	Sep 2016	$100K
Implement new processes, procedures, and technologies (via pilot studies of new techniques and technologies) that will result in enhanced design approval processes.	Research Report—Implementation of modifications to the approval process	Mar 2017	$500K
Monitor and collect data on new processes, procedures, and technologies.	Dashboard and reports	Sep 2017	$200K
Validate measures and results of new processes, procedures, and technologies.	Research Report—Validation of design certification process modifications	Jan 2018	$200K
Produce a final report with results, findings, and suggestions for recommended modifications of processes, procedures, and applicable	Research Report—Recommendations for modification	Feb 2018	$100K

technologies that can be integrated into the ATO workforce.	to approval process and develop opportunities to work collaboratively with industry.		
Disseminate to ATO and support development of implementation plans.	Discussions with ATO	Sep 2018	$200K

7. RESOURCE REQUIREMENTS

It is estimated that the 5-year research effort will require two in-house technical full-time equivalents (FTE) and a half programmatic FTE for the duration. It is also estimated that research contractor support will be required in the amount of $4.6M over the 5-year research effort as detailed by the research tasks identified in Section 6.

Below is the 5-year funding profile requested for the entire research effort.

Fiscal Year	FAA	Contractor
2013	2.5 FTE	$200K
2014	2.5 FTE	$1.1M
2015	2.5 FTE	$500K
2016	2.5 FTE	$1.6M
2017	2.5 FTE	$700K
2018	2.5 FTE	$500K
Total		$4.6M

8. REFERENCES

1. FAA Modernization and Reform Act of 2012, Pub. L. No. 112-95, 126 Stat. 11 (February 14, 2012).

2. FAA Acquisition Management System, available at "http://fast.faa.gov"

3. FAA JO 1000.37, "Air Traffic Organization Safety Management System," March 19, 2007.

4. FAA AVS, "AVS Work Plan for NextGen 2012," March 2012

B

Committee and Staff Biographical Information

S. MICHAEL HUDSON, *Co-Chair*, is currently chairman and CEO of I Power Energy Systems, LLC. Prior to that he was vice chairman, Rolls-Royce North America Holdings, a position he assumed in early 2000 and continued until his retirement in spring 2002. He also held the positions of president and CEO, chief operating officer, and chief financial officer with Rolls-Royce Allison, following its acquisition by Rolls-Royce in 1995. He served on the boards of several joint venture companies in which Rolls-Royce Allison had interest. Mr. Hudson began his career with Pratt and Whitney Aircraft, working in aircraft engine design, installation, and performance, engine development and demonstration, and industrial and marine engine application engineering. Mr. Hudson is a fellow of the Society of Automotive Engineers and the Royal Aeronautical Society, an honorary fellow of the American Helicopter Society (AHS), and an associate fellow of the American Institute of Aeronautics and Astronautics (AIAA). Mr. Hudson served on the AIAA Propulsion Committee, the AHS Propulsion Committee, the board of directors of the National Association of Manufacturers, and the board of directors of Indianapolis Water Company, and he was chairman of the AHS board of directors. Mr. Hudson was a member of the Society of Automotive Engineers (SAE), and he served as chairman of the SAE's Aerospace Council and on its Aerospace Program Office and Finance Committees. He received the SAE Franklin W. Kolk Air Transportation Progress Award and the Royal Aeronautical Society British Gold Medal. Publications range from technical work on propulsion to defense procurement and business initiatives. Mr. Hudson served on Air Force and Department of Defense review groups, and he was a member of NASA's Aeronautics Advisory Committee and the Subcommittee on Rotorcraft Technology and chaired the Propulsion Aeronautics Research and Technology Subcommittee. He holds a B.S. in aeronautical engineering from the University of Texas, Austin. He has served on several National Research Council (NRC) Committees, including the Committee on Aeronautics Research and Technology for Environmental Compatibility, the Committee on Analysis of Air Force Engine Efficiency Improvement Options for Large Non-Fighter Aircraft, and the Committee on Materials Needs and R&D Strategy for Future Military Aerospace Propulsion Systems and has been a member of the NRC's Aeronautics and Space Engineering Board.

WILLIAM S. LEBER, JR., *Co-Chair*, is senior vice president of Air Traffic Innovations for PASSUR Aerospace. His duties include strategy formation and strategic alliances with other companies, universities, and research organizations. He was formerly a research analyst principal and senior manager for Lockheed Martin in business development for their Collaborative Air Traffic Management Practice where he coordinated Lockheed Martin's efforts in airport collaborative decision making and other collaborative air traffic management domains. He has

25 years of air traffic management experience coordinating with the Federal Aviation Administration (FAA) and other air navigation service providers in the Atlantic and Pacific regions. He was a chief flight dispatcher and worked for Northwest Airlines and Delta Air Lines for more than 26 years. Mr. Leber is a member of the FAA Research Engineering and Development Advisory Committee (REDAC)-National Airspace System Operations Subcommittee, where he was co-chair of the Weather-Air Traffic Management Integration Work Group. He is a former chair of the Collaborative Decision Making Future Concepts Working Group and was co-chair of the Air Transportation Association's overall collaborative decision making effort from 2001 to 2004. He is a former president and co-founder of the Airline Dispatchers Federation, a non-union professional association. He holds a B.S. in aeronautical administration from St. Louis University and holds aircraft dispatcher and pilot certificates.

JANDRIA S. ALEXANDER is the principal director of the Cyber Security Subdivision in the Engineering and Technology Group at the Aerospace Corporation. The subdivision activities focus on the technical development and delivery of cyber and information assurance (IA) capability in the Engineering and Technology Group. This includes internal and external coordination of cyber and IA test beds and laboratories, developing and enhancing technical capabilities, and supporting cyber and IA-related research and customer tasking. Ms. Alexander currently leads the engineering support for cyber, network, and information security services across the defense, intelligence community, civil, and commercial sectors. She leads teams performing systems engineering for cyber operations, including architecture, requirements, and concept of operations (CONOPS) support for integrating cyber operations into advanced ground and space segments. She also leads the cyber research on cyber operations, malware, space cyber, vulnerability analysis, wireless, mobility, cloud, enterprise architectures, and network infrastructure and monitoring. Ms. Alexander previously supported several customers leading independent assessments with cross-organizational federally funded research and development center (FFRDC) teams, performing strategic and tactical planning support and cybersecurity resiliency risk management, policy, and governance. Ms. Alexander supports several internal aerospace cyber activities: Cyber Corporate Strategic Initiative and the Corporate Research Thrust in Space Cyber. Prior to joining the Aerospace Corporation, Ms. Alexander led security engineering and research in key management, COMSEC software, penetration testing, operating system and application level security; security architectures, threats, vulnerabilities and countermeasures; and risk management for government and commercial customers. She is a commissioner on the Virginia State Cyber Security Commission. Ms. Alexander earned her B.S. in computer science from Brandeis University and M.S. in technology of management from American University. She currently serves on the NRC's Committee on Future Research Goals and Directions for Foundational Science in Cybersecurity.

STEVEN J. BROWN is the chief operating officer for the National Business Aviation Association. He is responsible for managing representation of the association's 10,000 members on all domestic and international aviation regulatory, safety, security, and operational policy issues. Mr. Brown previously served as the associate administrator for air traffic services at the FAA where he was responsible for the daily operation of the U.S. Air Traffic Control system. He led a workforce of more than 35,000 employees, from air traffic controllers to maintenance technicians, and administered a $9 billion annual budget. During his tenure, dozens of modernization and technology programs were either initiated or completed, the Y2K transition was accomplished and the recovery from the 9/11 attacks was successfully completed. Mr. Brown is a current commercial pilot and has previously worked as a full-time pilot and instructor. He also is currently serving as chairperson of the board of trustees for the Aviation Accreditation Board International, which is the specialized accrediting organization for university degree-based aviation programs worldwide. He is a graduate of executive development programs at Penn State University and the University of Southern California. Mr. Brown has earned both a B.A. in education and M.Ed. from Texas A&M University.

VICTORIA COX is a consultant for the Victoria Cox Solutions, LLC. Previously she served as the FAA's assistant administrator for the Next Generation Air Transportation System (NextGen). At FAA, she led the transformation of the nation's air traffic control system with responsibility for the multi-billion dollar NextGen portfolio. Within the FAA, Ms. Cox also served as the director of the Air Traffic Organization's International Office, the director of Flight Services Finance and Planning, and the program director of the Aviation Research Division. Prior to joining the FAA, she was director of International Technology Programs in the Office of the Director of Defense Research and Engineering, Office of the Secretary of Defense. A physicist, Ms. Cox served as chief of physics and scientific director of the European Office of Aerospace Research and Development in London. As senior scientist of the Seeker Evaluation Branch at the Wright Armament Laboratory, Eglin Air Force Base, she created and managed an organization that provided the then unique capability to generate synthetic signatures and measure performance of space-based sensors for the Ballistic Missile Defense effort. As a NASA scientist, she conducted research in space environmental effects and was responsible for thermal vacuum conditioning and testing of the Hubble Space Telescope. Since retiring from the FAA in 2013, she has served on the Executive Steering Committee for the AIAA's Aviation 2014 Conference and has recently initiated a consulting practice. She has a certificate in U.S. National Security Policy from Georgetown University. She also earned her private pilot's license in 1985. She earned a B.A. from Converse College and M.A.Ed. from East Carolina University.

JOSEPH M. DEL BALZO is the president and CEO of JDA Aviation Technology Solutions, a company offering a wide range of airport and airspace planning, safety, security, training, and technology application support to international civil aviation clients. As a direct result of his efforts, JDA has been formally recognized by the FAA to assist air carrier applicants in preparing for the FAR Part 121 air carrier certification process. He has extensive experience in leading safety audits, helping clients move new and innovative technology from the laboratory into the aviation system, successfully guiding new air carrier startups and new start up aircraft manufacturers through the FAA certification process, guiding Part 91 corporate aircraft operators through the FAA Part 135 certification process, and directing the development and implementation of safety management systems. He serves as an independent technical advisor to industry and government clients. He served as acting administrator and deputy administrator of the FAA and also served as FAA's executive director of system operations, an organization responsible for defining requirements for new technology; installing, operating and maintaining all air traffic control systems and facilities; operating the nation's air traffic control system; and developing and overseeing safety regulations for all aircraft, airline, and airmen in the National Airspace System. As FAA executive director for system development, Mr. Del Balzo developed long-range research and development programs to support the timely introduction of new technology into the U.S. air traffic control system; refocused FAA initiatives to increase capacity at U.S. airports; and established close working relationships with the members of the U.S. aviation community in developing system requirements for the 21st century. He served as director of the FAA Eastern Region responsible for installing, maintaining, and operating all air traffic control facilities in seven eastern states; overseeing safety compliance of all aircraft, airline, and airmen operating in the seven-state region and ensuring the safe operation of all commercial and general aviation airports located in the region. He is a fellow of the AIAA, a former member of the FAA REDAC, and former chairman of the Aircraft Safety Subcommittee, a former member of the board of directors and former chairman of the Air Traffic Control Association, and a former member of the Civil Tilt-Rotor Advisory Committee. Mr. Del Balzo was awarded an honorary doctor of science from Embry-Riddle Aeronautical University. He holds a B.S. in engineering from Manhattan College and an M.S. in engineering from Drexel University.

R. JOHN HANSMAN is the T. Wilson Professor of Aeronautics and Astronautics at the Massachusetts Institute of Technology (MIT), where he is also the director of the MIT International Center for Air Transportation. He conducts research in the application of information technology in operational aerospace systems. Dr. Hansman holds seven patents and has authored more than 250 technical publications. He has more than 5,800 hours of pilot in-command time in airplanes, helicopters, and sailplanes, including meteorological, production, and engineering flight test experience. Dr. Hansman chairs the FAA REDAC as well as other national and international advisory committees. He is co-director of the Aviation Sustainability Center, which is a multi-university FAA Center of Excellence. He is a member of the U.S. National Academy of Engineering and a fellow of the AIAA. He has received numerous awards, including the AIAA Dryden Lectureship in Aeronautics Research, the ATCA Kriske Air Traffic Award, a Laurel from *Aviation Week and Space Technology*, and the FAA Excellence in Aviation Award. He earned his Ph.D. in physics, aeronautics, and meteorology from MIT. He has served on the NRC's Committee on Review of the Enterprise Architecture, Software Development Approach, and Safety and Human Factor Design of the Next Generation Air Transportation System.

AMY R. PRITCHETT is the David S. Lewis Associate Professor of Cognitive Engineering in the School of Aerospace Engineering, Georgia Institute of Technology. She holds a joint appointment in the School of Industrial and Systems Engineering. Dr. Pritchett has led numerous research projects sponsored by industry, NASA, and the FAA. She has also served as director of NASA's Aviation Safety Program, responsible for planning and execution of the program, conducted at four NASA research centers and sponsoring roughly 200 research agreements; in that role, she also served on the Office of Science and Technology Policy Aeronautic Science and Technology Subcommittee and the executive committees of the Commercial Aviation Safety Team and the Aviation Safety Information Analysis and Sharing program. She has published more than 170 scholarly publications in conference proceedings and in scholarly journals such as *Human Factors*, the *Journal of Aircraft*, and *Air Traffic Control Quarterly*. She has also won the Radio Technical Commission for Aviation's William H. Jackson Award and, as part of Commercial Aviation Safety Team, the Collier Trophy. The AIAA has named a scholarship for her. Dr. Pritchett is the editor in chief of the *Journal of Cognitive Engineering and Decision Making*. She is also a licensed pilot. Dr. Pritchett received B.S., M.S., and Sc.D. degrees in aeronautics and astronautics from MIT. She recently chaired the NRC study examining air traffic controller staffing.

AGAM N. SINHA is currently the president of ANS Aviation International, LLC, providing aviation consulting to a wide base of organizations. Dr. Sinha retired from MITRE Corporation where he was a senior vice president as well as general manager of the Center for Advanced Aviation System Development (CAASD). He also directed the FAA's FFRDC. CAASD supports the FAA, Transportation Security Administration, and international civil aviation authorities in addressing operational and technical challenges to meet aviation's capacity, efficiency, safety, and security needs. Dr. Sinha has more than 40 years of experience in aviation and weather systems. He serves on the board of trustees of Vaughn College of Aeronautics in New York and is on the advisory board of Ph.D. in aviation at Embry Riddle Aeronautical University. He served as a member of the FAA's NextGen Advisory Committee and on the FAA REDAC. He was chairman of RTCA board of directors and an elected member of the RTCA Policy Board, Air Traffic Management Advisory Committee, and the Air Traffic Management Steering Group. He has also served on the Advisory Committee of Lincoln Lab at MIT and of the National Center of Atmospheric Research (Research Applications Programs). He is an associate fellow of the AIAA. He has more than 80 publications and has been an invited presenter to a wide range of organizations nationally and internationally. Dr. Sinha is the recipient of several awards and citations from the FAA and industry. He earned his Ph.D. in operations research from the University of Minnesota. He is a member the NRC's Aeronautical and Space Engineering Board and has served as chair of the Aviation Group of the Transportation Research Board.

EDMOND L. SOLIDAY is a state representative of the Indiana State Assembly. Previously, he served as vice president of safety, quality assurance, and security at United Airlines before his retirement. Additionally, he was a line-qualified pilot at United for 35 years. He has qualified to fly numerous civilian and military aircraft. He is a Vietnam War veteran, serving as an attack helicopter pilot. He has served on numerous aviation safety-related advisory boards and commissions, including the Gore Commission's Aviation Security Baseline Working Group; co-chair of the Commercial Aviation Safety Team; chairman of the Flight Operations Quality Assurance Advisory Rulemaking Committee; past chair of the Air Transport Association (ATA) Safety Council, ATA Environmental Committee, ATA Executive Sub-Committee, International Air Transport Association Flight Safety Committee, and Star Alliance Safety Committee; the executive board of the Flight Safety Foundation; and member of the NASA Aviation Safety Program Executive Panel and the MIT Global Airline Industry Program Advisory Group. Additionally, he served as adjunct professor of Aviation Safety and Security at George Washington University from 1999 through 2007 and as a member of the Adler Planetarium Board of Trustees. He has served as an aviation consultant to the Rand Corporation, Boeing Company, Greenbriar Equity, LLP, Skadden, Arp, Meagher and Flom, LLP, Quirk and Bachelor, PC, and Condon and Forsyth, LLP. Among his awards are the Bendix Trophy for Aviation Safety, the Vanguard Trophy, and the Laura Tabor Barbour International Air Safety Award, Federal Bureau of Investigation and FAA Distinguished Service Awards, the Distinguished Flying Cross, two Bronze Stars, and the Purple Heart. The Commercial Aviation Safety Team, which Captain Soliday was the founding co-chair, received the prestigious Collier Trophy for reducing the commercial aviation fatal accident rate by 83% in 10 years and for producing the longest period without a commercial aviation fatality in the history of the industry. He has served on the NRC's Sub-Committee on Transportation Security Technology, the Decadal Survey of NASA Aeronautics Research, the Complex Integrated Systems Panel (chair), the committee to review NASA aeronautics safety research programs, as well as several other NRC studies and peer reviews, including terms as a member of the NRC's Aeronautics and Space Engineering Board.

RAYMOND VALEIKA is an independent consultant, advising major companies in aviation and technical matters among airlines, original equipment manufacturers, private equity, and lessors. He is an internationally recognized aviation operations executive with over 40 years of managing large airline maintenance operations. He retired from Delta Airlines as senior vice president of technical operations. At Delta Airlines, Mr. Valeika directed a worldwide maintenance and engineering staff of more than 10,000 professionals, maintaining a fleet of nearly 600 aircraft. During his tenure, he oversaw the creation of Delta Technical Operations as a maintenance, repair, and operations (MRO) entity, which has grown to become one of the leading MRO service providers in the aviation industry. Through his leadership and focus on continuous improvement of the human processes in aviation maintenance, Delta Technical Operations consistently rated at the top of the industry for performance benchmarks in the areas of safety, quality, productivity, and reliability. In 2008, there were major issues with airworthiness directives implementation which resulted in major flight disruptions; Mr. Valeika was asked by the FAA to participate in a review of airworthiness as part of a special Airworthiness Directives Compliance Review Team. This effort resulted in recommending improved processes for the future of the FAA and industry. He earned his B.S. in aeronautical engineering from St. Louis University.

EDWARD L. WRIGHT is a professor of physics and astronomy at the department of physics and astronomy at the University of California, Los Angeles (UCLA). At UCLA, Dr. Wright has been the data team leader on the Cosmic Background Explorer (COBE), a co-investigator on the Wilkinson Microwave Anisotropy Probe (WMAP), an interdisciplinary scientist on the Spitzer Space Telescope, and the principal investigator on the Wide-field Infrared Survey Explorer (WISE). Dr. Wright is a member of the National Academy of Sciences. He is well known for his Cosmology Tutorial website for the informed public, and his web-based cosmology calculator for professional astronomers. He has recently served on the NRC's Beyond Einstein Program Assessment Committee, the committee to study Autonomy Research in Civil Aviation, and the committee to study NASA's planned WFIRST-Astrophysics Focused Telescope Assets (WFIRST-AFTA) program.

Staff

DWAYNE A. DAY, *Study Director,* a senior program officer for the NRC's Aeronautics and Space Engineering Board (ASEB), has a Ph.D. in political science from the George Washington University. Dr. Day joined the NRC as a program officer for the Space Studies Board (SSB). Before this, he served as an investigator for the Columbia Accident Investigation Board, was on the staff of the Congressional Budget Office, and also worked for the Space Policy Institute at the George Washington University. He has held Guggenheim and Verville fellowships and was an associate editor of the German spaceflight magazine *Raumfahrt Concrete,* in addition to writing for such publications as *Novosti Kosmonavtiki* (Russia), *Spaceflight,* and guest editing *Space Chronicle* (United Kingdom). He has served as study director for numerous NRC reports, including *3-D Printing in Space* (2014), *NASA's Strategic Direction and the Need for a National Consensus* (2012), *Continuing Kepler's Quest: Assessing Air Force Space Command's Astrodynamics Standards* (2012), *Preparing for the High Frontier: The Role and Training of NASA Astronauts in the Post-Space Shuttle Era* (2011), *Vision and Voyages for Planetary Science in the Decade 2013-2022* (2011), *Defending Planet Earth: Near Earth Object Surveys and Hazard Mitigation Strategies* (2010), *Opening*

New Frontiers in Space: Choices for the Next New Frontiers Announcement of Opportunity (2008), and *Grading NASA's Solar System Exploration Program: A Midterm Review* (2008).

MICHAEL MOLONEY is the director for Space and Aeronautics at the SSB and the ASEB of the National Research Council of the U.S. National Academies. Since joining the ASEB/SSB, Dr. Moloney has overseen the production of more than 40 reports, including four decadal surveys—in astronomy and astrophysics, planetary science, life and microgravity science, and solar and space physics—a review of the goals and direction of the U.S. human exploration program, a prioritization of NASA space technology roadmaps, as well as reports on issues such as NASA's Strategic Direction, orbital debris, the future of NASA's astronaut corps, and NASA's flight research program. Before joining the SSB and ASEB in 2010, Dr. Moloney was associate director of the BPA and study director for the decadal survey for astronomy and astrophysics (Astro2010). Since joining the NRC in 2001, Dr. Moloney has served as a study director at the National Materials Advisory Board, the Board on Physics and Astronomy (BPA), the Board on Manufacturing and Engineering Design, and the Center for Economic, Governance, and International Studies. Dr. Moloney has served as study director or senior staff for a series of reports on subject matters as varied as quantum physics, nanotechnology, cosmology, the operation of the nation's helium reserve, new anti-counterfeiting technologies for currency, corrosion science, and nuclear fusion. In addition to his professional experience at the National Academies, Dr. Moloney has more than 7 years' experience as a foreign-service officer for the Irish government—including serving at the Irish Embassy in Washington and the Irish Mission to the United Nations in New York. A physicist, Dr. Moloney did his Ph.D. work at Trinity College Dublin in Ireland. He received his undergraduate degree in experimental physics at University College Dublin, where he was awarded the Nevin Medal for Physics.

ANDREA M. REBHOLZ, program coordinator, joined the ASEB in 2009. She began her career at the National Academies in 2005 as a senior program assistant for the Institute of Medicine's Forum on Drug Discovery, Development, and Translation. Prior to the Academies, she worked in the communications department of a D.C.-based think tank. Ms. Rebholz graduated from George Mason University's New Century College with a B.A. in integrative studies-event management. She earned the Certified Meeting Professional (CMP) designation in 2013 and has more than 11 years of experience in event planning.